An *Introductory* Guide to DIFFERENTIAL CALCULUS

A FIRST COURSE IN DIFFERENTIATION

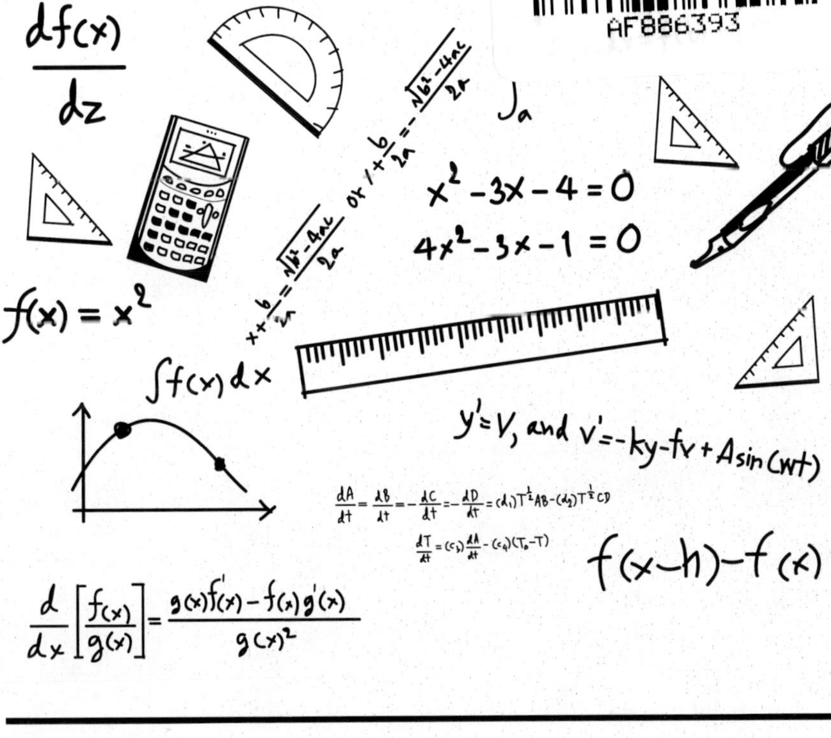

HENGCHEN ZHANG

An Introductory Guide To Differential Calculus
First Edition August 2024

First Published 2024
Reprinted 2024

ISBN: 978-1-923163-82-9 (Paperback)

NATIONAL LIBRARY OF AUSTRALIA

A catalogue record for this book is available from the National Library of Australia

Published by Hengchen Zhang and Clark & Mackay

Proudly printed in Australia by Clark & Mackay

ABOUT THIS BOOK

Differential calculus, the two words, are only simple reflections of the depth of the course content and what the subject matter has to offer, and not intended to be a term that scares students. The primary objective of this book is to introduce those with a genuine curiosity for mathematics to a new branch called calculus, which itself branches even further into differential and integral calculus. This book will focus on the former and its applications.

In order to facilitate interest in such students, the course material from various textbooks has been reorganised and rewritten as an introductory text, intended to teach the student the fundamentals of differential calculus. Most chapters have a set of exercises, some with real-world applications, to try and aid with making the abstract concept more accessible and tangible. Care has been exercised to avoid the abuse of notation, although it may still be present a couple of times in this book due to a lack of better communication choices.

As this book is meant as an introductory guide, concepts have been organised and presented relatively dumbed down and one-at-a-time so that the student can comprehend the nature of the questions clearly without having multiple ideas confusing the text's understandability. Knowledge of all pre-calculus mathematics, up to and including limits, is required to understand differentiation thoroughly, and it is highly advised that limits are well-understood or mastered before commencing this course. I highly recommend Dennis Donovan and David Bock's *Barron's AP Calculus Premium* for further reading and even more elaborate concepts that are not covered in this book.

The material presented includes a variety of tables, graphs, and charts, all of which are personally designed and manifested, unless specifically noted otherwise (see List of Figures). It is in the best interest of the student that they also familiarise themselves with introductory analysis, as the latter theorems will require some understanding of the topic. Differential calculus tends to converge towards the 'pure mathematics' side, rather than engineering, which is more physical and applied in comparison.

On the other hand, most of the problems in this book will be rounded to 2-3 decimal places or the same number of significant figures in the question. The universal constant π is to be taken to 10 significant figures (3.141592654), whenever radians are used instead of degrees. Similarly, rounding of decimals should be done in the last step, and not earlier, to maintain the preferred accuracy of the answer, unless specified otherwise.

Lastly, as one who has experienced the daunting feeling after beginning a differential calculus course, I can only offer some words of advice to help you on your journey ahead.

From Lao Tzu:

千里之行，始于足下

Translated, it means that *even journeys of a thousand miles begin with one step*.

When approaching questions or concepts that are too difficult or convoluted, just remember what you have achieved, and remember this saying. Always do not be afraid to take the first step, as is the case for some exercises in this book, because success will follow inevitably. This was true for myself, and will undoubtedly be true for everyone who works with passion and determination.

CONTENTS

PART IV: Concavity, Differentials, and Antidifferentiation

LIST OF FIGURES

PREFACE

I still remember the first time hearing and searching up the term 'calculus' all the way back in third grade. I recall the mass confusion of seeing letters pop up in maths for the first time, as well as functions, Greek letters and weird lines that did not make any sense. Now, here I am, writing a guide to differential calculus. It is also quite ironic how a love for maths and a love for writing culminates in a monstrosity of a block of text, numbers, and algorithms, quite conveniently named an 'introductory guide', that could frighten a student, especially one who is eager to learn about the intricacies and the wonderful world of differential calculus.

This book, nonetheless, is designed to prepare students, armed with sophisticated knowledge around limits, for the world of derivatives and its applications, leading up to the topics of concavity, differentials, and eventually briefly touching upon the notion of integral calculus. I trust that students with an inherent curiosity for these topics will grasp them with ease, and therefore excel through their respective calculus and analysis courses in practice. I make no claim to originality or excellence in the manifestation of the textbook problems, although care has been exercised to make them as realistic or as simplified as possible (in most situations) for the ease of understanding.

Unfortunately, this introductory guide will stop short of in-depth content in integral calculus, although the basics of antidifferentiation will be taught; those who wish to continue with it should be encouraged to do so with respectable textbooks. This book will also skip over the majority of the mathematics regarding limits, as this is expected to be well-understood, if not mastered, before beginning differential calculus, as should the topics of trigonometry, geometry, and algebra. The proper limit definition of the derivative, however, will be used and explained thoroughly in this guide as it remains the foundation for all calculus, differential and integral, as well as further mathematical applications of both, including but not limited to linear algebra, discrete mathematics, differential equations, and multivariable calculus.

Finally, it would not be surprising if errors were present few and far between this guide, as per the norm with textbooks – hopefully they remain mostly insignificant and do not corrupt the integrity of the content within. If any errors and/or misprints were found, it would be greatly appreciated if they were brought to my attention.

H. ZHANG
4 November 2023

For Mr. Peter Coyne.

PART I

Introduction to Differential Calculus

1.1. Introduction to Derivatives

Calculus deals with *infinitesimal change*. This infinitesimal change could be either infinitely big or small. Differential calculus, as the name suggests, deals with the latter. Differential calculus refers to the branch of calculus that involves *derivatives* and *instantaneous change*, which has many real-life applications in statistics, economics, and other related fields.

Simply, the derivative is the instantaneous *gradient* of a *function*, or the slope of the tangent line at a point with respect to any variable. This variable, more often than not, is x, although other variables are also used to differentiate, such as t (time) and θ (angle). Differentiation with these variables will appear later in this book. For now, however, let us stay with the standard x.

We have all seen this formula below – it is the average slope of a function between two points.

$$\frac{\Delta y}{\Delta x} = \frac{y_2 - y_1}{x_2 - x_1}$$

Given $(0, -4)$ and $(3, 2)$, we can find the average slope between these two points as follows:

$$\frac{\Delta y}{\Delta x} = \frac{2 - (-4)}{3 - 0} = 2$$

Therefore, the average slope of the secant line between the points $(0, -4)$ and $(3, 2)$ is 2. Finding the slope using this method is perfectly fine for linear functions, but it may be problematic for non-linear relationships, which do not always exhibit a constant gradient across their domain.

For example, take the simple quadratic function $y = x^2$ at the points $(0, 0)$, $(1, 1)$, and $(2, 4)$. At $x = 0$, the slope is 0; between $x = 0$ and $x = 1$, the average slope is 1. However, between $x = 1$ and $x = 2$, the slope of the secant line is 3. It is evident that the gradient changes across the function's domain, but how can we calculate this? More so, how can we calculate the instantaneous rate of change at any point?

We will learn how to compute derivatives in the next chapter but let us first familiarise ourselves with the notation. The most common method of representing the derivative is the Leibniz notation, $\frac{dy}{dx}$, named after German polymath Gottfried Wilhelm Leibniz, who co-invented calculus with Isaac Newton. His method of notation is very similar to that of the average slope formula, only replacing Δ with d.

Here, d is a *differential*, representing an infinitely small change (so small that it is negligible). The operator used to derive functions, logically, is $\frac{d}{dx}$, and for a function $f(x)$, the notation is $\frac{df}{dx}$ (and vice versa for others). Lagrange notation, named after Italian mathematician Joseph-Louis Lagrange, is also seen quite regularly: for a function $f(x)$, the derivative is $f'(x)$. Leibniz and Lagrange notation each have their pros and cons and will be used interchangeably in this book.

Newtonian notation, where the derivative of y is taken to be $\dot{y}$, is rarely seen in introductory calculus and will be omitted from this guide.

1.2. Limits and the Limit Definition

Leibniz notation introduces us to the differential d, meaning an infinitesimally small change. We can represent this using limits, where the interval between two x-values, h, approaches 0. Then, the gradient of the secant line between the two values approaches the derivative. A primitive definition for the derivative, notated using limits, can be seen below:

$$f'(a) = \lim_{x \to a} \frac{f(x) - f(a)}{x - a}$$

Despite its intuition, this form is rarely seen in differential calculus because it lacks practicality. The equation requires two points to compute an answer, which is inefficient and becomes difficult when dealing with extremely large or extremely small numbers. It should not be used in serious computations, only a tool to help us better understand the concept of a derivative.

A much more useful definition can be seen below:

$$f'(a) = \lim_{h \to 0} \frac{f(a+h) - f(a)}{h}$$

This definition recognises that, as h approaches 0, the distance between the two y-values $f(a+h)$ and $f(a)$ decreases until it is negligible. In theory, as the values $f(a+h)$ and $f(a)$ approach each other, the average slope converges towards the derivative. Through the notation of a limit, it thereby gives us our instantaneous gradient without needing to physically divide by 0. There is a technicality, however, in the sense that we are still dividing by 0, just algebraically: that is the entire idea surrounding limits, and what the value is merely approaching, rather than what is practically achievable.

Nonetheless, we arrive at the proper limit definition for the derivative, also known as the first principle of differentiation.

For a function $y = f(x)$, the derivative is given as follows:

$$\boxed{\frac{dy}{dx} = \lim_{h \to 0} \frac{f(x+h) - f(x)}{h}}$$

Hence, for a function $f(x) = 2x + 3$, we can substitute into our limit definition:

$$\begin{aligned}
\frac{dy}{dx} &= \lim_{h \to 0} \frac{2(x+h) + 3 - (2x+3)}{h} \\
&= \lim_{h \to 0} \frac{2x + 2h + 3 - (2x+3)}{h} \\
&= \lim_{h \to 0} \frac{2x + 3 + 2h - 2x - 3}{h} \\
&= \lim_{h \to 0} \frac{2h}{h} \\
&= 2
\end{aligned}$$

Therefore, the slope at any value of x over the entire domain of the function $f(x) = 2x + 3$ is 2. This, of course, can be calculated through the average slope formula because it is a linear relationship between x and y. We can use the first principles definition on non-linear functions to precisely calculate the derivative, which would be impossible otherwise.

Let $y = f(x)$, and $f(x) = x^2 + 5$. Then:

$$\begin{aligned} f'(x) &= \lim_{h \to 0} \frac{(x+h)^2 + 5 - (x^2 + 5)}{h} \\ &= \lim_{h \to 0} \frac{x^2 + 2xh + h^2 - x^2}{h} \\ &= \lim_{h \to 0} \frac{h(2x+h)}{h} \\ &= 2x \end{aligned}$$

Hence, we have deduced that, over the entire domain of the function f, the derivative is $2x$. Given the points $(-1, 1)$, $(0, 0)$, $(2, 4)$ and $(3, 9)$, we can easily find the slope at these points.

$$\begin{aligned} f'(-1) &= 2(-1) \\ &= -2 \\ f'(0) &= 2(0) \\ &= 0 \\ f'(2) &= 2(2) \\ &= 4 \\ f'(3) &= 2(3) \\ &= 9 \end{aligned}$$

As calculated, the derivatives of $f(x)$ at $x = \{-1, 0, 2, 3\}$ are $\{-2, 0, 4, 9\}$ respectively. This process of differentiation using limits can be tedious at times, especially trigonometric functions (since they often require complex identities), so more practical methods can be viewed in the following chapters.

However, before we move onto the more interesting derivative rules, we must recognise some caveats of differentiation:

1. Differentiability implies continuity. A function f can only be differentiable at $x = a$ if $\lim_{x \to a^-} f(x) = \lim_{x \to a^+} f(x)$, and $\lim_{x \to a} f(x) = f(a)$. This means that the function must approach $f(a)$ from both the negative and positive sides when $f(a)$ also exists (i.e. if there was a point discontinuity at $f(a)$, then the function is not differentiable at $x = a$). This also applies for jump and asymptotic discontinuities, although the limit does not exist and thus the derivative is undefined as well.
2. The derivative of a horizontal line is 0, but the derivative of a vertical line is undefined (when differentiating with respect to x). This is because, to calculate the slope between two points with the same x, but different y-values, you will need to divide by zero. Some may interpret this as a derivative of infinity, with the assumption of $\frac{\Delta y}{0} = \infty$; this is plausible but should not be a considered definition.
3. A function can only be differentiable at $x = a$ if $f(a)$ is not a sharp point on a graph. In Figure 1.a. (over the page), while the polynomial function x^{1000} appears to take a sharp turn at $x = -1$ and 1, when dilated in enough, still exhibits a curve-like contour. For the absolute value (modulus) function g (Fig. 1.b.), the two individual functions that make up the piecewise function, $g(x) = x$ and $g(x) = -x$, with different derivatives, contradict at $x = 0$ and therefore cannot be differentiated at that point.

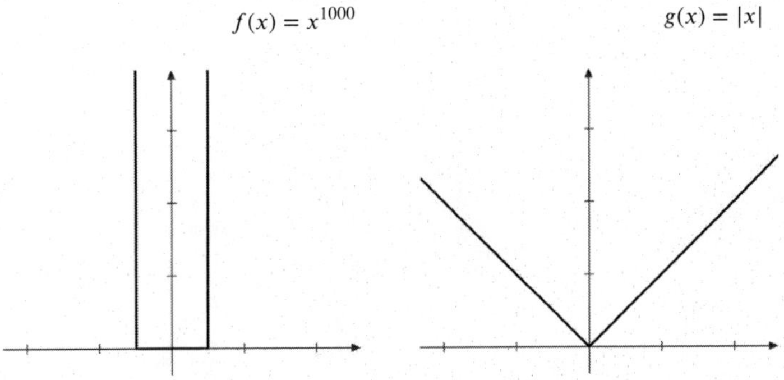

a. *b.*

Figure 1.

1.3. Differentiating Polynomials and the Power Rule

While we have already established the differentiation process, using first principles to find derivatives can be quite tedious, especially when tasked with higher-order functions. The power rule is the most fundamental rule in differentiation, and we can use it to compute the derivative of functions with numerical indices (e.g. polynomials). The rule is given below, applicable for all values of n: $n \in \mathbb{R}$:

$$\boxed{\frac{d}{dx}x^n = nx^{n-1}}$$

For example, given the functions $f(x) = x^2$ and $h(x) = x^3$, we can calculate the derivatives as follows:

$$
\begin{aligned}
\frac{df}{dx} &= \frac{d}{dx}\left[x^2\right] \\
&= 2 \cdot x^{2-1} \\
&= 2x \\
\frac{dh}{dx} &= \frac{d}{dx}\left[x^3\right] \\
&= 3x^2
\end{aligned}
$$

There are also some derivative properties regarding the process of addition and subtraction which are integral to understanding the differentiation of polynomials, listed below.

1. $\frac{d}{dx}[f(x) + g(x)] = f'(x) + g'(x)$
2. $\frac{d}{dx}[f(x) - g(x)] = f'(x) - g'(x)$
3. $\frac{d}{dx}[C \cdot f(x)] = C \cdot f'(x),\ C$ is a constant
4. $\frac{d}{dx}[C] = 0$

(The d and dx do not cancel as they represent separate differentials.)

The sum rule states that the derivative of the sum (of functions) is equal to the sum of the derivatives.

The difference rule states that the derivative of the difference is equal to the difference of the derivatives.

The constant multiplication rule states that the derivative of a constant multiplied by the function is equal to the constant multiplied by the derivative.

Finally, the constant rule states that the derivative of a constant is equal to 0.

It is helpful to think that the derivative operator distributes over addition and subtraction, but not over multiplication and division. All these derivative rules are, in one way or another, verified by limits and will be useful for differentiating polynomials and other functions with numerical indices.

For example, let $f(x) = 3x^3 + 4x^2 - 8x + 9$, then:

$$\begin{aligned} f'(x) &= \frac{d}{dx}\left[3x^3 + 4x^2 - 8x + 9\right] \\ &= \frac{d}{dx}\left[3x^3\right] + \frac{d}{dx}\left[4x^2\right] - \frac{d}{dx}[8x] + \frac{d}{dx}[9] \\ &= 3\frac{d}{dx}\left[x^3\right] + 4\frac{d}{dx}\left[x^2\right] - 8\frac{d}{dx}[x] + 0 \\ &= 3 \cdot 3x^2 + 4 \cdot 2x - 8 \\ &= 9x^2 + 8x - 8 \end{aligned}$$

While this seems complicated initially, the process is marginally sped up after repeated exercise over extended periods of time. Some exercises for differentiating polynomials are on the next page. Answers to these problems and general notes are available at the end of this book, on page 127.

1.3. Exercises

1. Find the derivative of the following functions **using the first principles of differentiation**:

a) $y = 3x + 2$

b) $y = 3x + 2x - 4$

c) $y = 10x - 5$

d) $y = 10x + 10$

e) $y = 2x - x$

f) $y = 22$

2. Find the derivative of the following functions **using the derivative rules**:

a) $y = x^2$

b) $y = 3x^3 - 2$

c) $y = -x^3$

d) $y = 12x^2 - 3x^3 + x$

e) $y = -2x^2 + 4x$

f) $y = x^5 + x^4 + x^3$

3. Find the derivatives of the following:

a) $y = x^{-1}$

b) $y = 3x^{-3} - 2x^{-2}$

c) $y = -x^{-3} + x^2$

d) $y = 2x^2 - 5x^{-3} + x^{-1}$

4. If differentiability implies continuity, does continuity imply differentiability? Prove or disprove this statement.

5. If $\frac{d}{dx}[x] = 1$, then why does $\frac{d}{dx}[1 + 1 + 1 + 1 \ldots]$ (x times) $= 0$?

1.4. The Product and Quotient Rules

While we have already established the derivative rules for adding and subtracting, the realm of functions extends far beyond those two basic operations. We can extend our knowledge of derivatives using the product and quotient rules, which allow us to differentiate functions involving multiplication and division respectively.

The product rule states, for a function $f(x) = g(x) \cdot h(x)$, then $f'(x) = g'(x) \cdot h(x) + g(x) \cdot h'(x)$. Often, we simplify $g(x) = u$ and $h(x) = v$, where:

$$\boxed{f'(x) = u'v + v'u}$$

This notation is more commonly used, where u and v also represent functions of the variable x.

Let $g(x) = (3x + 2)(5x^2)$. Instead of needing to multiply the polynomials and using the power rule, we can use the product rule here instead:

$$\begin{aligned} g(x) &= (3x + 2)\left(5x^2\right) \\ g'(x) &= \frac{d}{dx}[(3x + 2)] \cdot \left(5x^2\right) + \frac{d}{dx}\left[\left(5x^2\right)\right] \cdot (3x + 2) \\ &= 3\left(5x^2\right) + 10x(3x + 2) \\ &= 30x^2 + 15x^2 + 20x \\ &= 45x^2 + 20x \end{aligned}$$

The product rule is especially useful in larger functions comprised of several functions multiplied together. Using the answer above, we can now differentiate the function $\phi(x) = (3x + 2)(5x^2)(4x^4)$. Firstly, let us adapt the product rule into a general form that is applicable for $\phi = uvw$.

Given ϕ in a form of $\phi(x) = uvw$, where $u = 3x + 2$, $v = 5x^2$, and $w = 4x^4$. We can firstly differentiate uv, then apply the product rule a second time. The resultant would be $\phi'(x) = \frac{d}{dx}[u'v + v'u]w + w'[u'v + v'u]$, with sincere apologies for the rapid abuse of notation.

Nonetheless:

$$\begin{aligned}
\phi(x) &= (3x+2)\left(5x^2\right)\left(4x^4\right) \\
\phi'(x) &= \frac{d}{dx}\left[(3x+2)\left(5x^2\right)\right] \cdot \left(4x^4\right) + \frac{d}{dx}\left[4x^4\right](3x+2)(5x^2) \\
&= \left(45x^2 + 20x\right)\left(4x^4\right) + \left(16x^3\right)(3x+2)\left(5x^2\right) \\
&= 180x^6 + 80x^5 + 16x^3\left(15x^3 + 10x^2\right) \\
&= 180x^6 + 80x^5 + 240x^6 + 160x^5 \\
&= 420x^6 + 240x^5
\end{aligned}$$

The product rule is extremely versatile and can be used in to differentiate other functions by simply rearranging them. For example, the function $g(x) = \frac{2x^2+5}{x}$ in this form cannot be differentiated using the product rule. However, rewriting it as $g(x) = (2x^2 + 5)(x^{-1})$ makes the function differentiable using the product rule, instead of requiring the quotient rule, which is objectively more time-consuming and error-inducing.

To solve for the derivative:

$$\begin{aligned}
g'(x) &= \frac{d}{dx}\left[2x^2 + 5\right] \cdot \left(x^{-1}\right) + \frac{d}{dx}\left[x^{-1}\right] \cdot \left(2x^2 + 5\right) \\
&= 4x\left(x^{-1}\right) + \left(-x^{-2}\right)\left(2x^2 + 5\right) \\
&= 4 + \left(-2 - 5x^{-2}\right) \\
&= -5x^{-2} + 2
\end{aligned}$$

Despite most rational functions (two functions divided) being able to be rewritten as a product with negative indices, some of them cannot and must stay in fraction form. This is where the quotient rule comes in.

The quotient rule is the inverse of the product rule, and states that, for a function $f(x) = \frac{g(x)}{h(x)}$, the derivative will be $f'(x) = \frac{g'(x)h(x) - h'(x)g(x)}{[h(x)]^2}$.

For $g(x) = u$ and $h(x) = v$, the adapted formula is as follows:

$$\boxed{f'(x) = \frac{u'v - v'u}{v^2}}$$

If we wish to differentiate $f(x) = \frac{3x^2}{2x^4}$ in this form, we no longer need to rearrange. We can simply use the product rule, where $u = 3x^2$, $u' = 6x$, $v = 2x^4$, and $v' = 8x^3$.

$$\begin{aligned} f'(x) &= \frac{6x\left(2x^4\right) - 8x^3\left(3x^2\right)}{(2x^4)^2} \\ &= \frac{12x^5 - 24x^5}{4x^8} \\ &= \frac{3 - 6}{x^3} \\ &= -\frac{3}{x^3} \end{aligned}$$

A proof of the quotient rule can be seen later in this book, in the applications of the chain rule (Chapter 1.6).

1.4. Exercises

1. Differentiate the following using the **product rule**:

a) $y = 3x(10x)$

b) $y = 5x^{-2}(12x^9)$

c) $y = 2x^{-1}(10x^2)$

d) $y = 10x(5x^4)$

2. Differentiate the following using the **quotient rule**:

a) $y = \frac{5}{x^3}$

b) $y = \frac{4x^{15}-3x^2}{2x^3}$

c) $y = \frac{2x+5}{x^4}$

d) $y = (14)(2x^2)^{-2}$

3. Try to begin proving the quotient rule using the power rule and the product rule. *Hint: define the numerator and denominator, and then substitute for the function afterwards.*

1.5. Differentiating Trigonometric Functions

Rates of change in trigonometric functions are some of the most interesting topics in differential calculus because they can be defined, in one way or another, by another trigonometric expression. Take the fundamental $\sin(x)$ and $\cos(x)$ functions, and you may notice something peculiar:

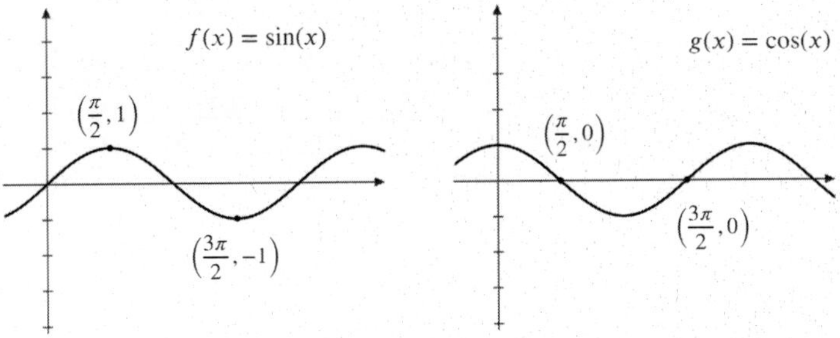

a. *b.*

Fig. 2.

Per Figure 2.a., we can see that the turning points (where $f'(x) = 0$) of $\sin(x)$ are at $\frac{\pi}{2}$ and $\frac{3\pi}{2}$, oscillating between 1 and -1 infinitely at intervals of π. In Figure 2.b., we see that $\cos(x) = 0$ at $x = \frac{\pi}{2}$ and $\frac{3\pi}{2}$, meaning that, when $f'(x) = 0$, $g(x) = 0$. If you have not concluded from inspection, $g(x) = f'(x)$, and therefore $\frac{d}{dx}\sin(x) = \cos(x)$.

Other 'coincidences' occur with some of the remaining trigonometric functions, and they can be verified using first principles and lots of trigonometric identities. These, however, will be omitted from this guide for 'introductory' purposes.

Nonetheless, we can still view and learn the relationship between several of these trigonometric functions and their derivatives, given on the table.

Original Function	Derivative
$\sin(x)$	$\cos(x)$
$\cos(x)$	$-\sin(x)$
$\tan(x)$	$\sec^2(x)$
$\sec(x)$	$\sec(x)\tan(x)$
$\csc(x)$	$-\csc(x)\cot(x)$
$\cot(x)$	$-\csc^2(x)$

Likewise, we can perform the product and quotient rules on the original functions (left).

$$\begin{aligned}\frac{d}{dx}[\sin(x)\cos(x)] &= (\cos(x))^2 - (\sin(x))^2 \\ &= \cos^2(x) - \sin^2(x)\end{aligned}$$

And:

$$\begin{aligned}\frac{d}{dx}\frac{\sin(x)}{\cos(x)} &= \frac{(\cos(x))^2 + (\sin(x))^2}{(\cos(x))^2} \\ &= 1 + \frac{\sin^2(x)}{\cos^2(x)} \\ &= \sec^2(x)\end{aligned}$$

It is noteworthy that, while $\frac{\sin(x)}{\cos(x)} = \tan(x)$, the derivative is equal to $\sec^2(x)$, the derivative of $\tan(x)$. It is very interesting and reassuring to compare the results obtained using differentiation principles against those manipulated through trigonometric identities.

However, it must also be noted that, for functions like $\sin^2(x)$, we cannot simply use the power rule and calculate the derivative to be $2\sin(x)$. To compute the derivatives of these, we must use the chain rule because they are composed functions (next chapter).

1.5. Exercises

1. Differentiate the following using the product rule:

a) $y = x\sin(x)$

b) $y = x^2\cos(x)$

c) $y = x^{-1}(\cos(x))$

d) $y = \sec(x)\tan(x)$

e) $y = -\csc(x)\cot(x)$

f) $y = 2\sin(x)\sec(x)$

2. Differentiate the following using the quotient rule:

a) $y = \frac{\sin(x)}{x}$

b) $y = \frac{\tan(x)}{2x^4}$

c) $y = \frac{7x^4}{\cos(x)}$

d) $y = \frac{\csc(x)}{\cot(x)}$

e) $y = \frac{15\sin(x)}{\cot(x)}$

f) $y = \frac{-\csc(x)\tan(x)}{\sin(x)\cot(x)}$

3. Why can we not differentiate $\sin^2(x)$ to be $2\sin(x)$? What other methods will be required? Does this apply to other trigonometric functions as well?

4. If the derivative of x^2 is $2x$, why does the answer to Question 3 apply?

1.6. The Chain Rule

As the name suggests, the chain rule is used to differentiate 'chained' functions. They refer to functions within functions, or alternatively known as 'composed' functions. We have already learnt to differentiate basic trigonometric functions and polynomials, but now we can attempt to differentiate more complex ones such as $\sin^2(x)$ and $(5x-3)^3$. The chain rule states, for a function $y = f\left(g(x)\right)$ or $(f \circ g)(x)$:

$$\boxed{\frac{dy}{dx} = \frac{df}{dg} \cdot \frac{dg}{dx} = f'\left(g(x)\right) \cdot g'(x)}$$

Note that f refers to the outer function, and g refers to the inner function. For example, for a function $y = (2x+3)^2$, f is the squared function, and g is $2x+3$. Especially in the instance of $\sin^2(x)$, notation may inhibit the visual inspection and recognition of inner and outer functions. This should be well-noted when approaching the chain rule.

We may be unfamiliar with the notation of $\frac{df}{dg}$, but by treating both f and g as x-terms, this becomes more intuitive.

For $f(x) = \sin^2(x)$, $\frac{df}{dg} = \frac{d(\sin^2(x))}{d(\sin(x))} = \frac{d}{d(\sin(x))}\left[\sin^2(x)\right]$, which equates to $2\sin(x)$ (see next page). With the chain rule, we being to treat differentials (df, dg, dh, dx etc.) as algebraic terms which can be multiplied, as long as you do not cancel the d themselves – that would be distasteful and utter sacrilege.

For instance:

$$\frac{d}{dx}[y] = \frac{dy}{dx}$$

Let a function $g(x) = \sin^2(x)$. We can find $g'(x)$ as follows:

$$g'(x) = \frac{d\left(\sin(x)^2\right)}{d(\sin(x))} \cdot \frac{d(\sin(x))}{dx}$$
$$= 2\sin(x)\cos(x)$$

For multiple functions composed together $f(g(h(x)))$, the chain rule can be manipulated as follows:

$\frac{dy}{dx} = \frac{df}{dg} \cdot \frac{dg}{dh} \cdot \frac{dh}{dx} = f'\left(g\left(h(x)\right)\right) \cdot g'\left(h(x)\right) \cdot h'(x)$... and vice versa for longer chains.

A common appearance of the chain rule can be seen in the proof of the quotient rule, where y is expressed as $\frac{f(x)}{g(x)}$. We can rewrite this as $f(x)$ multiplied by $\frac{1}{g(x)}$. Then, we can use the product rule in combination with the chain rule to prove:

$$y = f(x) \cdot \frac{1}{g(x)}$$
$$y' = f'(x) \cdot \frac{1}{g(x)} + f(x) \cdot \frac{d}{dx}\frac{1}{g(x)}$$

We can now differentiate $\frac{1}{g(x)}$ using the chain rule:

$$= \frac{f'(x)}{g(x)} + f(x) \cdot -\frac{1}{\left(g(x)\right)^2} \cdot g'(x)$$
$$= \frac{f'(x)g(x)}{\left(g(x)\right)^2} - \frac{f(x)g'(x)}{\left(g(x)\right)^2}$$
$$= \frac{f'(x)g(x) - f(x)g'(x)}{\left(g(x)\right)^2}$$
$$= \frac{u'v - v'u}{v^2} \text{ as required. } \blacksquare$$

(The thick vertical stroke, sometimes called the tombstone, is often implemented to denote the end of the proof, similar to Q.E.D.)

1.6. Exercises

1. Differentiate the following using the chain rule:

a) $y = \cos^2(x)$ b) $y = \sin^3(x)$
c) $y = (4x+3)^4$ d) $y = \sqrt{34x^2 - 24}$
e) $y = \sqrt[3]{8x+2}$ f) $y = 2\sin^2(x)\cos(x)$

2. For the following, find $\frac{dy}{dx}$:

a) $y = -2(1+x)^4$ b) $y = (2x^2+5x+1)^3$
c) $y = (2x+3)^6$ d) $y = 2\sin(3\cos(4x))$

3. Differentiate the following function R with **respect to t.** *Hint: try and express the x-variable as a function of time.*

$$R(x) = 4x^4 + 7x^2 - 5x + 9$$

4. **Extension**: Assume the radius is a function of time. Find $\frac{dV}{dt}$. Note: this is in the domain of related rates, which will be covered later in this guide. Leave the answer with derivative notation on the opposite side of the equal sign as it is not a given value.

$$V = \frac{4}{3}\pi r^3$$

1.7. Differentiating Exponential and Logarithmic Functions

Exponential and logarithmic functions exhibit some of the most peculiar behaviours in mathematics. Take the constant e for example, it has several definitions, including:

$$e = \lim_{n\to\infty}\left(1+\frac{1}{n}\right)^n = \lim_{h\to 0}\left(1+\frac{h}{x}\right)^{\frac{x}{h}}$$

Note on the latter definition: x/h is similar to $1/n$ as both of them tend towards 0, while x/h can be regarded as n, for both of them have unbounded limits.

Euler's number can also be expressed as a summation of terms, as shown below:

$$e = \sum_{n=0}^{\infty}\frac{1}{n!} \approx 2.718$$

Similarly, e also has interesting properties in the field of calculus, such as the derivative, where:

$$\boxed{\frac{d}{dx}e^x = e^x}$$

For logarithms with base e (natural logarithms):

$$\begin{aligned}\frac{d}{dx}\log_e(x) &= \frac{d}{dx}\ln(x)\\ &= \frac{1}{x}\end{aligned}$$

These two properties of e are essential when approaching questions involving the differentiation of exponential and logarithmic functions.

A formal proof for the derivative of $\ln(x)$ is as follows. Take into consideration the differentiation of x with respect to y.

Let $y = \ln(x)$:

$$\begin{aligned} x &= e^y \\ \frac{d}{dy}x &= \frac{d}{dy}e^y \\ \frac{dx}{dy} &= e^y \\ \frac{dy}{dx} &= \frac{1}{e^y} \\ \frac{dy}{dx} &= \frac{1}{x} \quad \blacksquare \end{aligned}$$

This proof simply changes the variable we differentiate (with respect to) to y, where the differentiation identity for e^y is implemented. By flipping $\frac{dx}{dy}$, we get our derivative to be $\frac{1}{x}$.

This is another application of algebraic manipulation with differentials. For exponents and logarithms with any base, we can also express these in the form of a natural exponential or logarithm, before using the necessary product and chain rules.

Let us denote an exponential with base a (in the general case). Then:

$$\begin{aligned} y &= a^x \\ &= \left(e^{\ln(a)}\right)^x \\ &= e^{x\ln(a)} \end{aligned}$$

We can now easily differentiate this rearranged form using our knowledge of the chain rule.

Taking the derivative of both sides:

$$\begin{aligned}\frac{dy}{dx} &= \frac{d}{dx}e^{x\ln(a)} \\ &= \ln(a)\cdot e^{x\ln(a)} \\ &= \ln(a)\cdot a^{x} \\ &= a^{x}\ln(a)\end{aligned}$$

Differentiating regular logarithms is slightly more complex as it requires some more algebraic manipulation:

$$\begin{aligned}\log_a(b) &= \frac{\log_c(b)}{\log_c(a)} \\ &= \frac{\ln(x)}{\ln(a)} \\ &= \frac{1}{\ln(a)}\cdot\ln(x)\end{aligned}$$

To differentiate:

$$\begin{aligned}\frac{d}{dx}\log_a(x) &= \frac{d}{dx}\left[\frac{1}{\ln(a)}\cdot\ln(x)\right] \\ &= \frac{1}{\ln(a)}\cdot\frac{d}{dx}\ln(x) \\ &= \frac{1}{x}\cdot\frac{1}{\ln(a)} \\ &= \frac{1}{x\ln(a)}\end{aligned}$$

(Note that $1/\ln(a)$ is a constant.)

1.7. Exercises

1. Differentiate the following:

a) $y = e^{-x}$ b) $y = 2a^{-x+2}$
c) $y = \log(2x^2)$ d) $y = \log_2(5x^3 + x^2 + 3x)$
e) $y = 2\ln(4x + 9)$ f) $y = e^{x^2}$

2. Differentiate the following:

a) $y = -7\log_4(x)$ b) $y = 5 \cdot 2^x$
c) $y = -3\log(x)$ d) $y = 10^{x^2-1}$
e) $y = \log_4(x^3 - 7x + 5)$

1.8 Derivatives of Inverse Functions

After taking a look at exponential and logarithmic functions, it is fitting to look at some other inverse functions and their derivatives.

By defining two functions, f and g, as inverses of each other, they exhibit the following properties:

$$1)\quad f(x) = g^{-1}(x)$$
$$2)\quad g\big(f(x)\big) = x$$

If we differentiate the second expression, we get the following through the chain rule:

$$\frac{d}{dx}g\big(f(x)\big) = \frac{d}{dx}x$$
$$g'\big(f(x)\big) \cdot f'(x) = 1$$

$$\boxed{f'(x) = \frac{1}{g'(f(x))}}$$

Given $f(x) = e^x$ and $f^{-1}(x) = g(x)$, we can now differentiate the inverse functions and test whether the above is true. Firstly, we know that $g(x) = \ln(x)$ because it is the inverse of e^x. We can now find f' and g' accordingly:

$$f'(x) = e^x$$
$$g'(x) = \frac{1}{x}$$

With f' and g', let us substitute into the formula:

$$f'(x) = \frac{1}{g'\left(f(x)\right)}$$
$$e^x = \frac{1}{\frac{1}{e^x}}$$

Therefore:

$$e^x = e^x$$

(Note: this is a trivial demonstration on how the formula works and its validity in practice, not necessarily a question you would see on an exam.)

Let us try another example with points rather than the actual functions.

Let h be the inverse of g, where $h(3) = 4$ and $g'(4) = 0.5$. Find $h'(3)$.

$$g^{-1}(x) = h(x)$$
$$h'(x) = \frac{1}{g'\left(h(x)\right)}$$
$$h'(3) = \frac{1}{g'\left(h(3)\right)}$$
$$= \frac{1}{g'(4)}$$
$$= \frac{1}{0.5}$$
$$= 2$$

Therefore, the derivative of $h(x)$ at $x = 3$ is 2.

Inverse functions also include inverse trigonometric functions. These will build upon our previous understanding of trigonometric functions as well as the derivatives of inverse functions.

For inverse sine (arcsine): Let $y = \arcsin(x)$, then $x = \sin(y)$:

$$\frac{d}{dx}\sin(y) = \frac{d}{dx}x$$
$$\cos(y) \cdot \frac{dy}{dx} = 1$$

Then:

$$\begin{aligned}\frac{dy}{dx} &= \frac{1}{\cos(y)} \\ &= \frac{1}{\sqrt{1-(\sin(y))^2}} \\ &= \frac{1}{\sqrt{1-x^2}}\end{aligned}$$

Note: by the Pythagorean identity $\sin^2(A) + \cos^2(A) = 1$, we can also evaluate $\cos(y)$ to be $\sqrt{1-(\sin(y))^2}$.

For inverse cosine (arccosine): Let $y = \arccos(x)$, then $x = \cos(y)$:

$$\frac{d}{dx}\cos(y) = \frac{d}{dx}x$$
$$-\sin(y) \cdot \frac{dy}{dx} = 1$$

Then:

$$\begin{aligned}\frac{dy}{dx} &= -\frac{1}{\sin(y)} \\ &= -\frac{1}{\sqrt{1-(\cos(y))^2}} \\ &= -\frac{1}{\sqrt{1-x^2}}\end{aligned}$$

It is quite interesting, I find, how the derivatives of inverse sine and inverse cosine only differ by a negative sign, just like the derivatives of regular sine and cosine. This is also the case for the hyperbolic sine and cosine functions ($\sinh(x)$, $\cosh(x)$), but they will not be covered further due to their complex nature.

For inverse tangent (arctangent): Let $y = \arctan(x)$, then $x = \tan(y)$:

$$\begin{aligned}\frac{d}{dx}\tan(y) &= \frac{d}{dx}x \\ \sec^2(y)\cdot\frac{dy}{dx} &= 1\end{aligned}$$

Then:

$$\begin{aligned}\frac{dy}{dx} &= \frac{1}{\sec^2(y)} \\ &= \frac{1}{1+\tan^2(y)} \\ &= \frac{1}{1+x^2}\end{aligned}$$

After these steps, we now know how to differentiate more complex exponential and logarithmic functions, and can apply the product, quotient, and chain rules as per usual in practice.

(Note: inverse trigonometric functions are only compatible with radians.)

1.8. Exercises

1. Differentiate the following:

a) $y = \arctan(2x)$
b) $y = \arcsin(e^x)$
c) $y = -4\arccos(3x^4 - 2x^2 + 1)$
d) $y = \log(\arcsin(x))$
e) $y = e^{\arctan(2x)}$
f) $y = 5^{\arccos(3x^2)}$

2. Let f be the inverse of g, where $f(7) = 9$ and $g'(9) = 2$. Find $f'(7)$.

3. For $f(x) = 3x^2 + 4x + 4$ and $f(x) = g^{-1}(x)$, find $g'(3)$ when $g(3) = 2$.

4. A function h and its derivative has been represented on the table for certain x-values. If g is the inverse of h, find $g'(6)$.

x	$g(x)$	$h(x)$	$h'(x)$
2	0	6	1/3
6	2	8	3/2

1.9. Higher Order Derivatives

As we progress through our knowledge on differential calculus, we will begin to note that certain variables are interrelated through differentiation (and integration). In physics, most motion equations are derived from one another, and motion graphs are common method in displaying this relationship.

Four variables form the fundamentals of physics: displacement (position), velocity, acceleration, and time. We can thereby construct three different motion graphs using these four quantities (Fig. 3.).

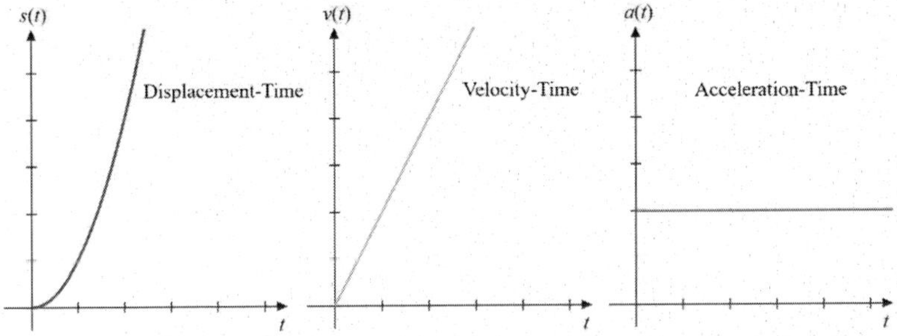

Fig. 3.

Theory and inspection tell us that the slope of the displacement-time graph is the velocity, and the slope of the velocity-time graph is acceleration. *Note: the integration process (which will be touched upon at the end of this book) is the reverse, where the area under the acceleration is the velocity, and the area under the velocity is the displacement.*

We can notate this as:

$$a(t) = v'(t)$$
$$v(t) = s'(t)$$

Therefore, acceleration is the derivative of the derivative of displacement. This 'derivative of a derivative' can be expressed as a **second derivative**, and its notation is given as:

$$\boxed{\frac{d}{dx}\frac{dy}{dx} = \frac{d^2y}{dx^2} = y''}$$

Note how the differentials again multiply to give the second derivative operator. We can now notate the relationship between acceleration, velocity, and displacement in a single line, as $a(t) = v'(t) = s''(t)$.

Note: the list of derivatives does not end at acceleration: the derivative of acceleration is the jerk, and the derivative of jerk is the snap. The latter two vector quantities, however, are rarely implemented and only apply for non-uniform acceleration (e.g. changing angles of inclination of a ramp).

For the function $f(x) = 3\sin(x) + 2x^2$, let us find its second derivative.

$$\begin{aligned} f'(x) &= \frac{d}{dx}3\sin(x) + \frac{d}{dx}2x^2 \\ &= 3\cos(x) + 4x \\ f''(x) &= \frac{d}{dx}3\cos(x) + \frac{d}{dx}4x \\ &= -3\sin(x) + 4 \end{aligned}$$

Therefore, the second derivative of $3\sin(x) + 2x^2$ is $-3\sin(x) + 4$.

1.9. Exercises

1. Find the second derivative of the following functions:

a) $y = 7x + 2$
b) $y = 3x + 2x - 4$
c) $y = 10x^3 - 5x$
d) $y = 10x + 10x^5$
e) $y = 2x - x^3$
f) $y = 19$

2. Find the second derivative of the following:

a) $y = -7\log_4(x)$
b) $y = 5 \cdot 2^x$
c) $y = -3\log(x)$
d) $y = 10^{x^2-1}$

3. Find other instances of second derivative applications.

4. How may we compute the velocity given acceleration? What does this mean and how can this be calculated algebraically?

PART II

Initial Applications of Derivatives

2.1. Implicit Differentiation

As our knowledge of differentiating functions becomes more sophisticated, our understanding of relations must not be negated either. Functions are a special type of relation, where each x corresponds to only one y. This can also be categorised as a one-to-one or many-to-one relationship between the two variables. A function and relation can be told apart using the vertical-line test.

If the vertical line only intersects the equation once for $\forall x$: $x \in \mathbb{R}$, it passes the test and can be categorised as a function. If the graph fails at any point, it is referred to as an implicit relation. We are used to $y = f(x)$ and therefore the notation $y' = f'(x)$, but implicit differentiation usually only requires the usage of Leibniz notation for clarity as well as more rigorous algebraic proofs for non-functions.

Figure 4.a. below denotes the function $f(x) = \frac{2x^2}{x}$, and it is clear that each x only corresponds to one $f(x)$. This applies over the whole domain of x except when $x = 0$, as the value is undefined, and the line is discontinuous. In other words, $f(x)$ is defined over the entire domain of $x, x \neq 0$. Fig. 4.b. shows the graph of a circle, a noteworthy relationship, which would fail the vertical line test.

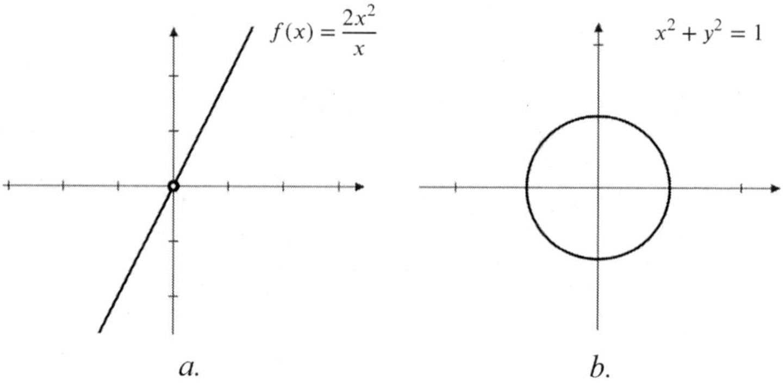

Fig. 4.

Notes: in Fig. 4.a., a removable (point) discontinuity can be viewed at the origin, this does not affect its status as a function; in Fig. 4.b., the derivative of the circle at $x = -1$ and 1 would be undefined as it is a vertical line (see Chapter 1.2: Limits and the Limit Definition).

Nonetheless, differentiation becomes more complicated when a relationship between two variables, x and y, is not expressed in the form $y = f(x)$. For example:

$$y^2 = 4x$$

If we try and square root the expression and then differentiate, we will get the undesired line:

$$y = \pm 2\sqrt{x}$$

This is a problem, both graphically and algebraically. Firstly, for each x, there is more than one y, and the slope of each line/curve may differ from the other without any apparent correlation. This is more evident graphically. Furthermore, the process of differentiation becomes more tedious with the added plus-minus sign, outputting two different answers, or being unsolvable in the worst case scenario.

However, we can combat both issues through implicit differentiation, where we simply take the derivative of both sides of the equation, then rearrange and factor out the derivative, $\frac{dy}{dx}$. Given a point on the graph, we can also find the instantaneous slope through substitution.

For the following relation, we can find the derivative using implicit differentiation, which requires taking the derivative with respect to x of both sides:

$$x^2 + y^2 = 1$$

$$\frac{d}{dx}\left[x^2 + y^2\right] = \frac{d}{dx}[1]$$

$$\frac{d}{dx}x^2 + \frac{d}{dx}y^2 = \frac{d}{dx}[1]$$

$$2x + 2y\frac{dy}{dx} = 0$$

$$2y\frac{dy}{dx} = -2x$$

$$\frac{dy}{dx} = -\frac{x}{y}$$

Therefore, the derivative of the relation is $-\frac{x}{y}$.

(Note: by the chain rule, the derivative of y^2 is $2y$ multiplied by dy/dx.)

For a more complicated relation:

$$
\begin{aligned}
y^4 + 3xy + x &= 2 \\
\frac{d}{dx}y^4 + \frac{d}{dx}3xy + \frac{d}{dx}x &= \frac{d}{dx}[2] \\
4y^3\frac{dy}{dx} + 3\left(y + x\frac{dy}{dx}\right) + 1 &= 0 \\
4y^3\frac{dy}{dx} + 3y + 3x\frac{dy}{dx} + 1 &= 0 \\
4y^3\frac{dy}{dx} + 3x\frac{dy}{dx} &= -3y - 1 \\
\frac{dy}{dx}\left(4y^3 + 3x\right) &= -(3y + 1) \\
\frac{dy}{dx} &= -\frac{3y+1}{4y^3+3x}
\end{aligned}
$$

(Note: in line 3, we used the product rule to differentiate $3xy$.)

If the question asks to find the derivative at the point $(2, 0)$ on the graph, we can simply substitute in the respective x and y to find the derivative.

$$
\begin{aligned}
\frac{dy}{dx} &= -\frac{3y+1}{4y^3+3x} \\
&= -\frac{3(0)+1}{4(0)^3+3(2)} \\
&= -\frac{1}{6}
\end{aligned}
$$

2.1. Exercises

1. Find the derivative, $\frac{dy}{dx}$, of the following equations:

   a) $x^2 - 3xy + y^2 = 1$
   b) $x^2 + xy + y^3 = 0$
   c) $x^3y - 2x^2 + y^4 = 8$

2. For 1.a), find the instantaneous slope of the slope at (1, 0).

3. For 1.b), find the derivative of the relation at the origin.

4. For 1.c), find the slope at $x = 0$. ($y \approx 1.68$, use this value to compute the derivative.)

2.2. Rates of Change

Calculus, both differential and integral, is all about rates of change. It is about time we have a proper discussion about the related formulae and notation regarding this topic, as well as its relevant nuances with regards to typical exam questions.

From thus forth, the layout of this book will encompass more of a textbook-style, with more in-depth exercises and considered formatting due to the sheer amount of information needed to develop a comprehensive understanding of these more advanced topics.

2.2.1. One-dimensional Motion

In previous chapters, we have touched upon one-dimensional motion as a topic deeply intertwined with differential calculus (as well as high school physics). We have recognised already that functions of displacement, velocity and acceleration (with respect to time) are related through differentiation, and the inverse through the integration process (covered at the end of this book).

RECALL

$$a(t) = v'(t) = s''(t)$$

Given this information, we can now add limitations to these functions and domains. For example, limiting t such that $t \in \mathbb{R}^+$ makes all values of t positive, which makes sense because time cannot be quantified negatively. In this, reducing the domain in question increases realism in problems, which also helps students understand the fundamentals of this topic.

One-dimensional motion can be interpreted as an object moving in a straight line and can be simulated by assigning values or functions for acceleration, velocity, and displacement.

Let us see an example of this below.

EXAMPLE 2.2.1.1.

Let a car's motion along a straight line be a function of time in the equation $x(t) = t^3 + 2t^2 - 5$, where x denotes Bob's displacement in metres, such that $t \geq 0$. Find his acceleration at $t = 2.5$ seconds.

SOLUTION

To find the general expression for the car's acceleration, we need to interpret the context: we are given $x(t)$, which is its displacement. Acceleration is the second derivative of displacement, meaning that we must differentiate $x(t)$ twice to get our general form for acceleration.

$$\begin{aligned} x(t) &= t^3 + 2t^2 - 5 \\ x'(t) &= 3t^2 + 4t \\ x''(t) &= 6t + 4 \end{aligned}$$

Since $x''(t)$ is the same as $a(t)$, the car's acceleration is $6t + 4$. Now, we simply substitute in $t = 2.5$ for our answer.

$$\begin{aligned} 6t + 4 &= 6(2.5) + 4 \\ &= 19\text{ms}^{-2} \end{aligned}$$

Therefore, the car is accelerating at 19ms^{-2} at $t = 2.5$ seconds.

2.2.2. Interpreting Direction in Motion

While time, t, cannot be negative (in most instances), the displacement, velocity, and acceleration certainly can be, depending on the function. A negative displacement simply means that the object or person has moved back a certain distance (in a one-dimensional simulation), a negative acceleration indicates that it is slowing down (deceleration). However, a negative velocity means that they have turned and are moving in the opposite direction. What is interesting about this property is that we can find where the object changes direction through simple mathematical analysis.

KEY POINT

An object changes direction when the value of velocity of the object changes signs (from positive to negative or vice versa) or crosses the x-axis graphically.

EXAMPLE 2.2.2.1.

Given the table to the right, find the interval(s) which the object changes direction.

Time (s)	Velocity (ms^{-1})
0	0
1	3
2	4
3	0
4	−1
5	2

a) $0 < t < 1$
b) $1 < t < 2$
c) $1 < t < 3$
d) $2 < t < 3; 0 < t < 1$
e) $3 < t < 4; 4 < t < 5$

SOLUTION

From the table, all values of velocity remain positive until $t = 4$, which has a negative velocity. A second later, at $t = 5$, the sign changes again back to positive. Therefore, the sign has changed twice, once between $t = 3$ and $t = 4$, the other between $t = 4$ and $t = 5$.
Hence, the correct answer is option e).

We can also analyse direction of motion, as well as changes in displacement and velocity through graphs. The contours of functions tell us more about the behaviour of the object than simple equations alone, where specific points on the lines and curves exhibit certain properties which can be revealed through differentiation. The following points should be noted:

KEY POINT

1. On a displacement-time (or position-time) graph, a positive derivative means that the object is moving forward, while a negative derivative means the opposite. A slope of 0 indicates that the object is stationary, and a y-value of 0 suggests that it has moved back to where it has started from (the original position).
2. On a velocity-time graph, a positive slope indicates that the object is speeding up, and vice versa. A flat slope indicates constant velocity, and a velocity of 0 means that it is stationary.
3. On an acceleration-time graph, a derivative of 0 means that the rate of acceleration is constant. A y-value of 0 either suggests constant velocity or no velocity (the object is stationary).

Let us take a look at Figure 5 below, a position-time graph:

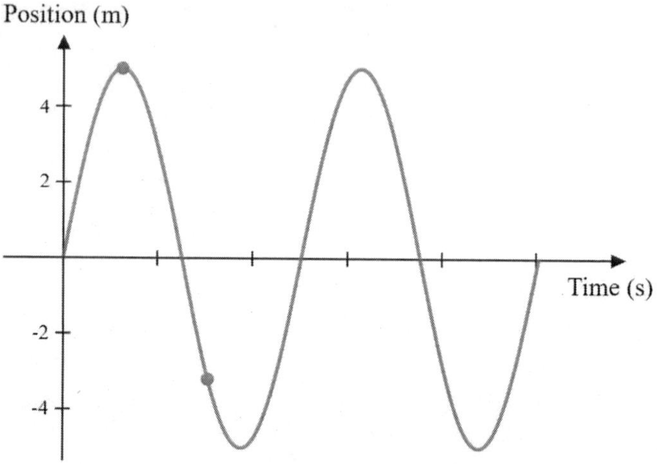

Fig. 5.

EXAMPLE 2.2.2.2.

Identify the two points on the function f in Figure 5, a displacement-time graph, and determine what it reveals about its velocity and acceleration respectively.

SOLUTION

This is a classic graphical question and asks for what the two points on the function suggest about the object's velocity (first derivative), acceleration (second derivative), and direction from graphical analysis. We will look at each point separately.

SOLUTION

For point 1 (left):
Let the x-value of the point be equal to a, while the y-value is equal to 5.
The instantaneous slope (first derivative) is zero (flat), indicating that the velocity is 0, and thereby meaning that the object is stationary. In addition, the first derivative decreases to zero as $y \rightarrow 5$, before it switches signs past this point. Hence, we can say that the object is about the change direction. The second derivative is a little more difficult to grasp conceptually, but we are able to conduct some rough assumptions.

In the first upside-down ∪-shaped curve, the derivative keeps decreasing until $x = 0$, before it gets increasingly positive*. Therefore, the curve of f' will be the steepest (in the negative direction) at $x = a$. If we conceptualise this, then f'' will have the most negative value at $x = a$, meaning the object is decelerating very quickly (this will be reviewed in Part IV).

For point 2 (right):
Let the x-value of the point be equal to b.
The function f has a negative derivative, meaning it has a negative velocity and is moving left. The rate at which the object is moving away from the origin is slowing down as x approaches the trough point, meaning that it is accelerating.

**Note:* *the term 'increasingly positive' simply means getting larger in value and can describe both positive and negative values, not exclusive to the former.*

Let us take a look at Figure 6 below, a velocity-time graph:

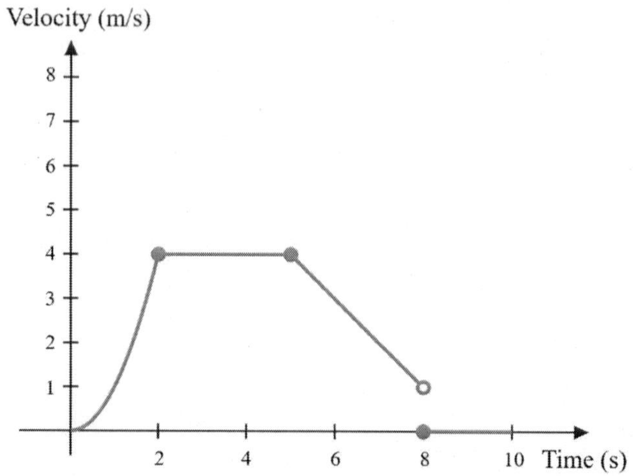

Fig. 6.

EXAMPLE 2.2.2.3.

Analyse the direction, velocity, and acceleration for the above piecewise function at $t = 1, 2, 4, 7$ and 8.

SOLUTION

At $t = 1$:
The velocity is 1ms^{-1}, observable from the graph. The derivative is positive therefore the object is accelerating. It is moving towards the right.

At $t = 2$:
The velocity is 4ms^{-1} but the derivative is undefined, therefore the acceleration cannot be determined.

SOLUTION

At $t = 4$:

The velocity is again 4ms^{-1} but the derivative is 0, indicating a constant velocity, and moving towards the right.

At $t = 7$:

The velocity is 2ms^{-1} with a negative derivative. The object is moving towards the left.

At $t = 8$:

The velocity is 0ms^{-1}, meaning that the object is stationary. The acceleration is undefined*.

**Note: the derivative is undefined at $x = 8$ because of the jump discontinuity.*

EXAMPLE 2.2.2.4.

A fish tank is leaking water at a non-uniform rate. In 20 hours, the 60L tank will have completely drained of all water.

The volume of water left in the tank after t hours is modelled by:

$$V(t) = 60(1 - 0.05t)^2 : t \in [0, 20]$$

Find the rate that the tank is losing water after five hours.

SOLUTION

We are given that $V(t) = 60(1 - 0.05t)^2$ models the volume of water left in the tank with a domain of $\{0 \leq t \leq 20\}$. We need to find the rate of change, and hence the first derivative of this given function. To do this, we can apply the chain rule:

$$\begin{aligned} V(t) &= 60(1 - 0.05t)^2 \\ V'(t) &= 60 \cdot 2(1 - 0.05t) \cdot -0.05 \\ &= -6(1 - 0.05t) \end{aligned}$$

Now, we can substitute in $t = 5$.

$$\begin{aligned} V'(t) &= -6(1 - 0.05t) \\ V'(5) &= -6(1 - 0.05 \cdot 5) \\ &= -6(0.75) \\ &= -4.5 \end{aligned}$$

Therefore, the rate that the tank is losing water after five hours is −4.5L/h.

2.2. Exercises

1. Refer to Figure 7 below. Describe, at the points A, B, and C, whether the object is accelerating, decelerating, or travelling at a constant velocity, including the direction of motion (assume right is positive).

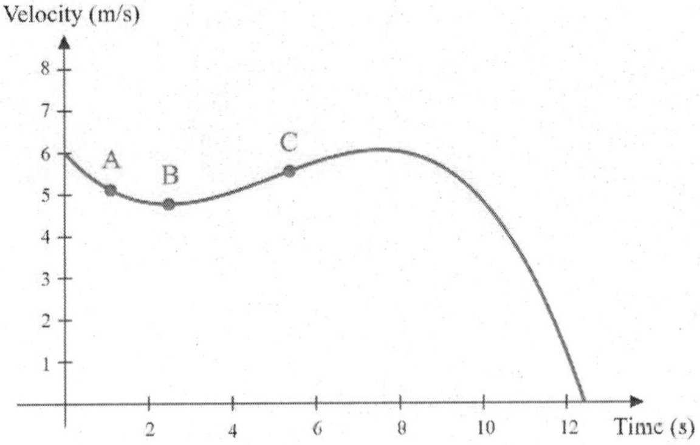

Fig. 7.

2. Jamie learnt 200 words in a vocabulary test. His forgetfulness can be represented by the equation $V(t) = 2(10 - x)^2$, where he forgets all the 200 words after 10 days. Find the rate of his forgetfulness after 2 days (assume $t \in \mathbb{Z}^+$).

3. **Extension**: A paper plane is thrown from the top of a hill towards the bottom. Its displacement can be represented by the equation $P(t) = -0.05(t - 10)(t - 15)(t - 20) + 25$. Find the instance(s) when the paper plane has a velocity of 0. Give your answer, t, to 4 significant figures.

4. The function $D(t) = \frac{50\sin(t)}{t}, t \in (0, 40]$ represents the distance between object A and object B up to 40 seconds. Find the rate of change of D when $t = 15$ seconds and $t = 25$ seconds.

5. **Extension**: In the SUVAT equations, $v = u + at$ is the derivative of $s = ut + \frac{1}{2}at^2$. Explain why the velocity equation is derived this way, and how come we do not apply the product rule on the ut and $\frac{1}{2}at^2$? What happens if we derive $v = u + at$ and are we able to keep finding the derivative to compute the jerk?

2.3 Related Rates

As an extension of implicit differentiation, chain rule applications, and rates of change, related rates deal with multi-dimensional change and motion (kinematics). Related rates problems are a culmination of previous, applied calculus knowledge, with given values for fixed rates to solve for the rate which the desired quantity is changing (e.g. rate of change of surface area given increasing dimensions, rate at which a spherical balloon is inflating etc.). This is usually done by redefining variables with respect to time and then relating them back to the question through differentiation. Most types of problems fit into two categories: trigonometry and related functions.

2.3.1 Trigonometry in Related Rates

Trigonometry itself is a broad, all-encompassing topic; this is no different in related rates. Possible questions include differentiating the Pythagorean Theorem, working with angles as functions etc. We will work through both cases below.

EXAMPLE 2.3.1.1.

Sam is walking North at a rate of $3ms^{-1}$ from a fixed point. Sam's friend David also starts from that point but walks East at a rate of $2ms^{-1}$. Find the rate at which they are moving away from each other after 10 seconds.

SOLUTION

Firstly, we need to define functions for Sam and David. The variable is not important in this step.

Let Sam's line of motion be abbreviated to *S*. Let David's motion be notated as *D*. Let the hypotenuse distance between *S* and *D* be equal to *H*.

Since Sam and David are moving exactly a right angle apart, the case fits the classic Pythagorean Theorem (see below).

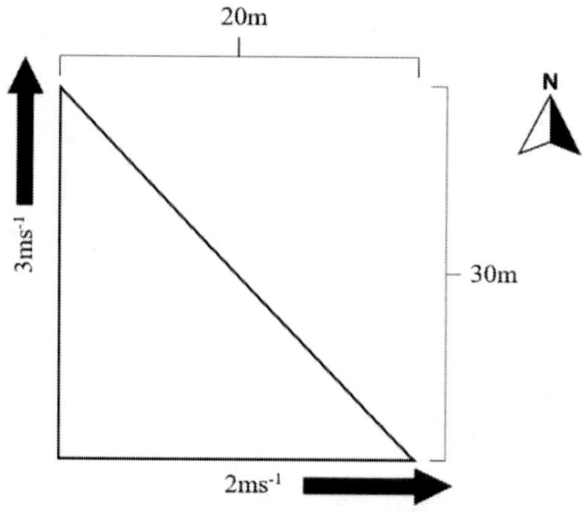

Fig. 8.

Substituting in our own variables:

$$S^2 + D^2 = H^2$$

Now, we differentiate by the independent variable time (t):

$$2S\frac{dS}{dt} + 2D\frac{dD}{dt} = 2H\frac{dH}{dt}$$

SOLUTION

Now, we can substitute our known variables into the equation. We know that $S = 30\text{m}$ and $D = 20\text{m}$. Our hypotenuse length hence will be equal to $\sqrt{1300}$ m.

After substitution:

$$2(30)\frac{dS}{dt} + 2(20)\frac{dD}{dt} = 2(\sqrt{1300})\frac{dH}{dt}$$

We actually know what $\frac{dS}{dt}$ and $\frac{dD}{dt}$ are from the question's context – they are simply the velocity at which Sam and David travel respectively (note that whether the velocity is positive or negative is respective to the point which they are approaching). We can now substitute and cancel the coefficient of 2 on either side:

$$30 \cdot 3 + 20 \cdot 2 = \sqrt{1300}\frac{dH}{dt}$$
$$\frac{dH}{dt} = \frac{130}{\sqrt{1300}}$$
$$= 3.606\text{ms}^{-1}$$

Therefore, the rate that Sam and David are moving away from each other at $t = 10$ is 3.606ms^{-1}.

Differentiating Pythagoras' Theorem is a common procedure in related rates, especially in motion (as we have just solved) and ladder problems (which are also conceptually similar). Within trigonometric ladder problems, we are subjected to more versatility with the variables – not necessarily just length and time, but also angles.

Let us take a look.

EXAMPLE 2.3.1.2.

A 15m long ladder is leaning on a wall. The perpendicular distance $x(t)$ from the bottom of the ladder to the wall is increasing at a rate of 2 metres per second. At $t = 0$, the ladder reaches 12m up the wall ($y(0) = 12$). What is the rate of change of the angle function $\theta(t)$ between the wall and the top of the ladder at that moment?

SOLUTION

The question has already defined the functions for the height of the wall and the length the ladder reaches on the floor. The ladder reaches 9m along the floor due to the shape formed being a 3-4-5 triangle. We also do not

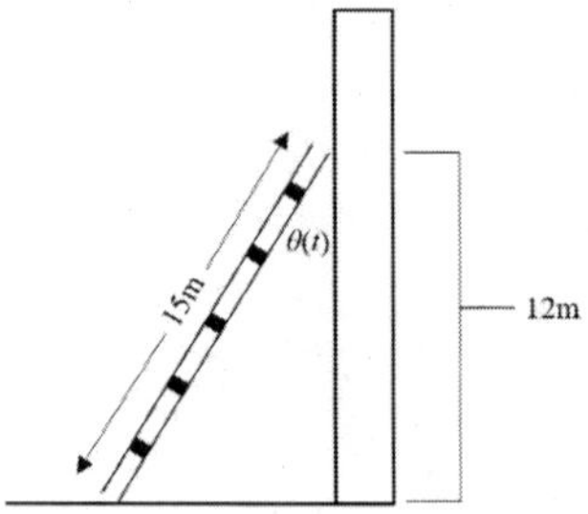

Fig. 9.

SOLUTION

need to define a function for the length of the ladder, L, as it remains at 15m, and hence L' will be equal to 0.
This case, unlike the previous, does not use Pythagoras' Theorem but proper trigonometry. We are given that x is increasing at a rate of 2ms^{-1}.

$$x'(t) = 2$$

We are also given that, at the beginning, the ladder reaches 12m up the wall: $y(0) = 12\text{m}$.

The sine of θ is the opposite divided by the hypotenuse, which is the distance from the bottom of the ladder to the wall (9m), and the length of the ladder (15m).

$$\sin(\theta) = \frac{opp}{hyp}$$

$$\sin\big(\theta(t)\big) = \frac{x(t)}{L}$$

$$x(t) = L \cdot \sin\big(\theta(t)\big)$$

Let us differentiate this to denote a function for $x'(t)$, which we already have an assigned value for.

$$x'(t) = L \cdot \cos\big(\theta(t)\big) \cdot \frac{d\theta}{dt}$$

Now, let us plug in the values for when $t = 0$.

SOLUTION

$$2 = 15 \cdot \cos\left(\arccos\left(\frac{4}{5}\right)\right) \cdot \frac{d\theta}{dt}$$

$$\frac{d\theta}{dt} = \frac{2}{15 \cdot 0.8}$$

$$= \frac{1}{6} \text{ radians per second}$$

$$= 9.55^\circ \text{ per second}$$

Therefore, the rate at which $\theta(t)$ is increasing at $t = 0$ is $\frac{1}{6}$ radians per second or 9.55 degrees per second. Note that trigonometric functions are defined in radians, and should the question ask for degrees, the value should be converted.

2.3.2. Related Functions

Related functions, as the name suggests, are a series of two or more functions (with different variables) interrelated through one or more series of equations. Then, the derivative of one of the interrelated functions is given, often require differentiation using time or another variable through the chain rule.

Given the accumulated knowledge previously of the chain rule, the notion of related functions should not be difficult to grasp.

RECALL

$$\frac{dy}{dt} = \frac{dy}{dx} \cdot \frac{dx}{dt}$$

Note: For practical purposes, all questions in this book on related functions will be derived using the variable t, although it is also likely that other variables (e.g. u) will appear in similar questions and contexts.

EXAMPLE 2.3.2.1.

The time functions x and y are differentiable and are interrelated by the following equation:

$$y = 2x^2$$

Given that $\frac{dx}{dt} = 16$:

a) Find $\frac{dy}{dt}$.

b) Find $\frac{dy}{dt}$ when $x = 5$.

SOLUTION

a) For $\frac{dy}{dt}$:

$$\begin{aligned} y &= 2x^2 \\ \frac{dy}{dt} &= \frac{dy}{dx} \cdot \frac{dx}{dt} \\ &= 4x \cdot 16 \\ &= 64x \end{aligned}$$

Therefore, the general expression for $\frac{dy}{dt}$ is $64x$.

b) For $x = 5$:

$$\begin{aligned} \frac{dy}{dt} &= 64x \\ &= 64 \cdot 5 \\ &= 320 \end{aligned}$$

Therefore, the derivative of y with respect to t at $x = 5$ is 320.

2.3. Exercises

1. Tim and Andrew are approaching the same point from the North and West respectively. Tim is approaching at a rate of 4ms^{-1} while Andrew is taking his time and walking at a pace of 1ms^{-1}. Given that Tim is 50m away from the given point and Andrew is 100m away, find the rate that they are approaching each other at that instant.

2. The time functions x and y are related by the equation $y = 4x^7 + 2x$. Find the slope of y with respect to time given that $\frac{dx}{dt}$ is equal to 4.

3. A 13m long ladder is leaning on a wall. The perpendicular distance $x(t)$ from the bottom of the ladder to the wall is increasing at a rate of 1.5 metres per second. At $t = 0$, the ladder reaches 12m up the wall ($y(0) = 12$). What is the rate of change of the angle function $\theta(t)$ between the ground and the bottom of the ladder at that moment?

4. Jack is 50m away from a fixed point and is walking West at a rate of 1ms^{-1} towards it. Jack's good friend Miles starts walking North from the fixed point a rate of 2ms^{-1}. Find the rate at which they are moving away from each other after 25 seconds.

5. The time functions x and y are related by the equation $y = 7x^3 + 2x^2 + 1$. Find $\frac{dy}{dt}$ given that $\frac{dx}{dt}$ is equal to 3.

2.4 Linear Approximations

Often, calculus involves interpreting values from a table or a graph to come to definite conclusions. However, occasionally, we need to approximate values based on the limited data we are given. Linear approximation (or local linearity) allows us to approximate the *y*-value given the gradient. By assuming that the function remains relatively linear between two points, we can derive a relatively close answer through algebraic manipulation of the gradient and points in the tangent line formula.

Newton's method for linear approximation is ambiguous in cases for some polynomials and has occasional shortcomings and will thus be omitted from this book.

EXAMPLE 2.4.1.

The function $f(x)$ is defined for all x: $x \in \mathbb{R}$. Given the following information, approximate $f(1.9)$.

$$f(2) = 1$$
$$f'(2) = 4$$
$$f''(2) = 2$$

SOLUTION

This is a typical question for linear approximations. We can adapt previously taught formulae to help us better understand the problem.

SOLUTION

As we are approximating using a linear function, the general form can be expressed as $y - y_1 = f' \cdot (x - x_1)$, which you may recognise as the tangent line formula.

Adapting the equation and substituting:

$$\begin{aligned} y - f(2) &= f'(2) \cdot (x - 2) \\ y - 1 &= 4(1.9 - 2) \\ y - 1 &= -0.4 \\ y &\approx 0.6 \end{aligned}$$

Hence, through linear approximation, $f(1.9)$ is approximately equal to 0.6.

Note: the second derivative of f at $x = 2$ was not needed.

2.4. Exercises

1. Given that $g(4) = 6$ and $g'(4) = -2$, approximate $g(4.2)$ assuming $x \in \mathbb{R}$.

2. Given that $h(-3) = 2$ and $h'(-3) = -5$, approximate $h(-2.9)$ assuming x is defined over the closed interval $[-5, 0]$.

2.5. L'Hôpital's Rule

Guillaume de l'Hôpital, back in 1696, published the first textbook on differential calculus. In his book, he states that the limit of the derivative is equal to the derivative of the limit, given both are defined as real numbers, and the limit is indeterminate through substitution.

(Note: L'Hôpital's rule is also known as L'Hospital's rule (American spelling) and Bernoulli's rule.)

KEY POINT

Given that $\lim_{x\to c}\frac{f(x)}{g(x)}$ and $\lim_{x\to c}\frac{f'(x)}{g'(x)} \in \mathbb{R}$:

$$\lim_{x\to c}\frac{f(x)}{g(x)} = \lim_{x\to c}\frac{f'(x)}{g'(x)}$$

Where:

$$\frac{f(c)}{g(c)} = \frac{0}{0} \text{ or } \pm\frac{\infty}{\infty}$$

L'Hôpital's rule allows us to calculate limits without the need for rigorous proofs or complex trigonometric identities. It also incorporates both knowledge on limits and differentiation, allowing us to practise both. Let us try to evaluate a simple indeterminate limit using L'Hôpital's rule.

EXAMPLE 2.5.1.

Evaluate the limit below:

$$\lim_{x\to 1}\frac{1-x^2}{\ln(x)}$$

SOLUTION

Substituting in $x = 1$ into the limit:

$$= \frac{1 - 1^2}{\ln(1)}$$
$$= \frac{0}{0}$$

As $\frac{0}{0}$ is indeterminate, we can now use L'Hôpital's rule.

Let us define $f(x) = 1 - x^2$ and $g(x) = \ln(x)$. Then:

$$f'(x) = \frac{d}{dx}\left[1 - x^2\right]$$
$$= -2x$$

And:

$$g'(x) = \frac{d}{dx}\ln(x)$$
$$= \frac{1}{x}$$

We now have:

$$\lim_{x \to 1} \frac{-2x}{\frac{1}{x}}$$
$$= \lim_{x \to 1} -2x^2$$
$$= -2(1)^2$$
$$= -2$$

Therefore, the limit was evaluated to be -2 as x approached 1.

Let us also try a limit with x approaching infinity (see below).

EXAMPLE 2.5.2.

Evaluate the limit below:

$$\lim_{x \to \infty} \frac{7x^2}{x - x^3}$$

SOLUTION

Substituting in $x = \infty$ into the limit:

$$\begin{aligned} &= \frac{7(\infty)^2}{\infty - (\infty)^3} \\ &= -\frac{\infty}{\infty} \end{aligned}$$

We again have an indeterminate form. For $f(x) = 7x^2$ and $g(x) = x - x^3$:

$$\begin{aligned} f'(x) &= \frac{d}{dx}\left[7x^2\right] \\ &= 14x \end{aligned}$$

$$\begin{aligned} g'(x) &= \frac{d}{dx}\left[x - x^3\right] \\ &= 1 - 3x^2 \end{aligned}$$

We now have:

$$\lim_{x \to \infty} \frac{14x}{1 - 3x^2}$$

SOLUTION

However, this still returns an indeterminate $\frac{\infty}{\infty}$. We can now apply L'Hôpital's rule again:

$$= \lim_{x \to \infty} \frac{14}{-6x}$$

And again:

$$= \lim_{x \to \infty} \frac{0}{-6}$$
$$= 0$$

Therefore, the limit tended to 0 as x approached infinity.

2.5. Exercises

1. Evaluate the following limits:

a) $\lim_{x \to 0} \frac{\tan(x)}{3x + \tan(x)}$

b) $\lim_{x \to 0} \frac{\sin(x)}{5x + 3\sin(x)}$

c) $\lim_{x \to \infty} \frac{4x^2 - 7x}{(\ln(x))^2}$

d) $\lim_{x \to \infty} \frac{-3x^3 + 2x}{4x + 8}$

e) $\lim_{x \to 1} \frac{x\ln(x)}{x^2 - 1}$

PART III

Further Applications of Derivatives

3.1. Mean Value Theorem

The Mean Value Theorem (MVT) articulates a special type of relationship between the derivative of a function, $\frac{dy}{dx}$, and the average slope of the secant line, $\frac{\Delta y}{\Delta x}$, between any two points $(a, f(a))$ and $(b, f(b))$. Both the average and instantaneous slopes were defined in Part I.

The MVT states that, for a function $f(x)$ defined, continuous, and differentiable over the closed interval $[a, b]$, then there exists some arbitrary point(s) given as $(c, f(c))$ such that $c \in [a, b]$, where the derivative, $f'(c)$, is equal to the slope of the function's secant line over $[a, b]$.

This is given as:

KEY POINT

$$\frac{\Delta y}{\Delta x} = \frac{f(b) - f(a)}{b - a} = f'(c)$$

A visual representation of the Mean Value Theorem can be seen in Figure 10 below.

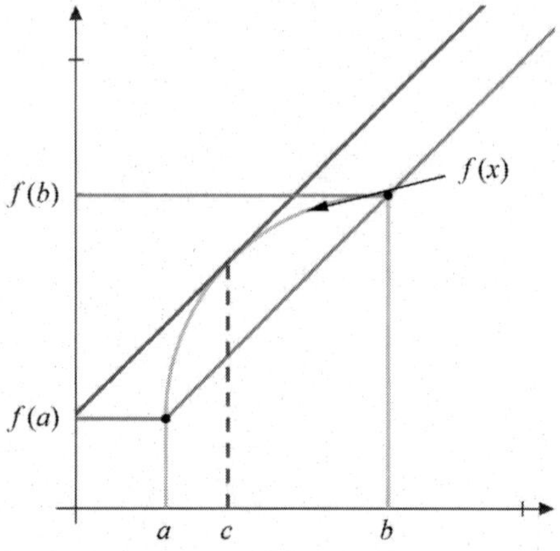

Fig. 10.

The MVT is useful with regards to analysing average values. For example, if a car's speed and average velocity were graphed with respect to time, the Mean Value Theorem could be used to indicate points which the car exceeded the speed limit, given quantities such as its distance travelled.

There are many proofs to the Mean Value Theorem, none of which are generally tested and hence will not be covered in this book, but we can still show that the theorem holds true.

EXAMPLE 3.1.1.

The function $f(x) = 4x^3 - 2x^2 - x$ is defined and continuous for $\forall x$: $x \in [2, 5]$. Find c, the x-value which has the same derivative as the average slope.

SOLUTION

Firstly, let us find the average slope of the function between $x = 2$ and $x = 5$.

$$\begin{aligned} f(x) &= 4x^3 - 2x^2 - x \\ f(5) &= 4(5)^3 - 2(5)^2 - 5 \\ &= 500 - 50 - 5 \\ &= 445 \\ f(2) &= 4(2)^3 - 2(2)^2 - 2 \\ &= 32 - 8 - 2 \\ &= 22 \end{aligned}$$

$$\begin{aligned} \frac{\Delta y}{\Delta x} &= \frac{f(b) - f(a)}{b - a} \\ &= \frac{445 - 22}{5 - 2} \\ &= \frac{423}{3} \\ &= 141 \end{aligned}$$

Now, we can differentiate the function.

$$\begin{aligned} f(x) &= 4x^3 - 2x^2 - x \\ f'(x) &= 12x^2 - 4x - 1 \end{aligned}$$

Substituting in the average slope:

$$\begin{aligned} 12x^2 - 4x - 1 &= 141 \\ 12x^2 - 4x - 142 &= 0 \\ 6x^2 - 2x - 71 &= 0 \end{aligned}$$

$$\begin{aligned} &x = 3.61, -3.28\text{: } c \in [2, 5] \\ &x = 3.61 \end{aligned}$$

Therefore, the point c occurs at $x = 3.61$.

3.2. Critical Points, Increasing and Decreasing Intervals

As we slowly introduce ourselves beyond the simple applications of differentiation, we begin to see the inherent patterns between the theoretical and graphical representations of functions and relations. Such is the case for the topics that will be covered in this section: critical points and types of intervals. Using these techniques, we can better analyse and interpret graphs and perceive them in real-life applications.

3.2.1. Critical Points

For a function $f(x)$ defined over its domain, then the derivative is given by $f'(x)$. A *critical point* or critical value, represented by the pronumeral c, relates to differentiability at $x = c$. Any existing c within the domain of f such that $f'(c) = 0$ or $f'(c)$ is undefined renders it a critical point of the function f. A function's derivative can only change signs from positive (increasing interval) to negative (decreasing interval), or vice versa, across critical points, but not always. We can find c by setting $f'(c)$ to 0, or by noting the values which f is undefined (known as the first derivative test). Then, we can create sign charts to find whether the derivative is positive or negative to analyse intervals. *Note: for a critical point to exist, the original function must be defined at that x-value.*

KEY POINT

A critical point c, is present on the function f when $f'(c) = 0$ or when $f'(c)$ is undefined.

Let us try an example below:

EXAMPLE 3.2.1.1.

Given $f(x) = 4x^3 + 6x^2$, find the critical points.

SOLUTION

As $f(x)$ is defined for all real numbers, critical points are only observable when $f'(c) = 0$.

$$f(x) = 4x^3 + 6x^2$$
$$f'(x) = 12x^2 + 12x$$

Substituting in $f'(x) = 0$:

$$12x^2 + 12x = 0$$
$$12x(x + 1) = 0$$

$$12x = 0, x + 1 = 0$$
$$x = -1, x = 0$$

Therefore, the critical points for $f(x)$ are at $x = -1, x = 0$.

EXAMPLE 3.2.1.2.

Find the critical points of $g(x) = x^5 - x^4$.

SOLUTION

Since $g(x)$ is defined for all real numbers, critical points are only observable when $f'(c) = 0$.

$$g(x) = x^5 - x^4$$
$$g'(x) = 5x^4 - 4x^3$$

Substituting in $g(x) = 0$:

$$5x^4 - 4x^3 = 0$$
$$x^3(5x - 4) = 0$$

$$x^3 = 0, 5x - 4 = 0$$
$$x = 0, x = \frac{4}{5}$$

Therefore, the critical points of $g(x)$ are at $x = 0$, $x = \frac{4}{5}$.

3.2.2. Extreme Value Theorem

The Extreme Value Theorem (EVT) is a theorem concerning function optimisation – the field of calculus concerning a graph's minimum and maximum points (and further applications).

The theorem states that for a function f continuous over the closed interval $[a, b]$, there exists both a minimum and maximum value of f over the interval. The theorem also states that there exists the x-values c and d within $[a, b]$, and that the case $f(c) \leq f(x) \leq f(d)$ holds true for $\forall x: x \in [a, b]$.

While this theorem may be almost self-explanatory, it remains an important notion that forms the foundation for all future work regarding intervals, extrema, and eventually concavity.

KEY POINT

> Let $f(x)$ be continuous over $[a, b]$:
> $\exists$ maximum value of f and minimum value of f;
> $\exists\, c, d \in [a, b]$ s.t. $f(c) \leq f(x) \leq f(d), \forall x: x \in [a, b]$.

3.2.3. Increasing and Decreasing Intervals

Let a function be defined as $f(x)$. The function f is increasing if and only if, over the minimum and maximum points in the interval, for all values of $x = a$ and $x = b$, then $a < b$ and $f(a) < f(b)$. On the contrary, f is said to be decreasing if, over the same interval, $a < b$ and $f(a) > f(b)$.

Of course, judging by the definition itself, this phenomenon is difficult to articulate using just mathematical language. We can recognise some of the knowledge and theorems listed previously in this guide and recall them for our benefit and use them to solve similar example problems.

For example, we can recall the nature of critical points and what they offer regarding optimisation.

RECALL

> c is present on the function f when $f'(c) = 0$ or when $f'(c)$ is undefined;
>
> $f'(x) < 0 \Longrightarrow$ function decreasing,
> $f'(x) > 0 \Longrightarrow$ function increasing.

Armed with sophisticated knowledge, let us try a problem:

EXAMPLE 3.2.2.1.

Given $f(x) = 2x^3 - 6x^2$, find the intervals which f is increasing.

SOLUTION

As $f(x)$ is defined for all real numbers, $f'(x)$ must also be defined across its domain, and critical points must only be observable when $f'(c) = 0$.

$$f(x) = 2x^3 - 6x^2$$
$$f'(x) = 6x^2 - 12x$$

Substituting in $f'(x) = 0$:

$$6x^2 - 12x = 0$$
$$6x(x - 2) = 0$$

$$6x = 0, x - 2 = 0$$
$$x = 0, x = 2$$

Therefore, the critical points for $f(x)$ are at $x = 0$, $x = 2$.

We can now notate these values on a number line, with the numbers between critical points as the unknown intervals. By substituting in values, we can find whether the derivative is positive (increasing) or negative (decreasing). The method is valid because a function's derivative, and hence the type of interval, can only change signs over a critical point. These processes are commonly known as 'interval calculations'.

SOLUTION

For $\{-\infty < x < 0\}$: (substitute in $x = -1$)

$$\begin{aligned} f'(x) &= 6x^2 - 12x \\ f'(-1) &= 6(-1)^2 - 12(-1) \\ &= 6 + 12 \\ &= 18\ (+ve) \end{aligned}$$

For $\{0 < x < 2\}$: (substitute in $x = 1$)

$$\begin{aligned} f'(x) &= 6x^2 - 12x \\ f'(1) &= 6(1)^2 - 12(1) \\ &= 6 - 12 \\ &= -6\ (-ve) \end{aligned}$$

For $\{2 < x < \infty\}$: (substitute in $x = 3$)

$$\begin{aligned} f'(x) &= 6x^2 - 12x \\ f'(3) &= 6(3)^2 - 12(3) \\ &= 54 - 36 \\ &= 18\ (+ve) \end{aligned}$$

From our calculations above, we get the following sign chart:

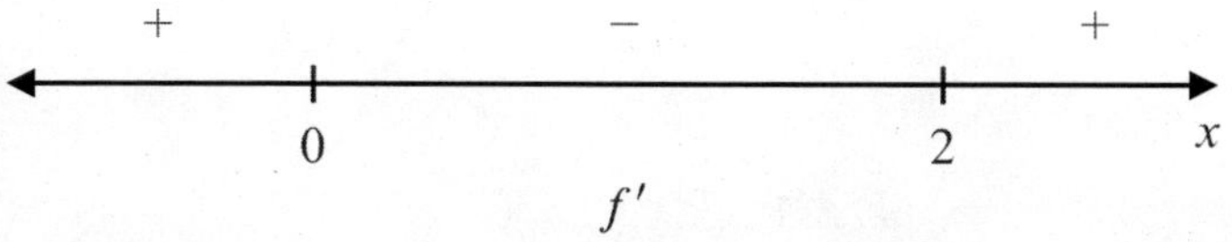

As a positive derivative indicates an increasing interval, f is increasing over the intervals $(-\infty, 0) \cup (2, \infty)$.

3.2. Exercises

1. Find all critical points of the function $f(x) = 5x^2 - 3x^5$.

2. Find all critical points of the function $f(x) = 9x^3 + 2x^7$.

3. Find all critical points of the function $h(x) = x^3 + x^{-1}$.

4. For the function in Question 1, find the intervals which f is increasing.

5. For the function in Question 2, find the intervals which f is decreasing.

6. For the function in Question 3, find the intervals which h is increasing.

3.3. Relative Extrema

When we talk about the turning point or the vertex of a parabola, we refer to the point at which the derivative of the function goes from being negative to positive, or vice versa. This is also commonly known as the maximum and minimum turning points, respectively, in pre-calculus courses.

What is valuable about the turning point on a parabola is that a second-degree polynomial is relatively simple to manipulate using completing the square, factorisation, and other methods. For example, given the standard form of the quadratic $y = ax^2 + bx + c$, the x-value of the turning point is given by $-\frac{b}{2a}$ and the y-value is equal to $-\frac{b^2}{4a}$. However, the process of finding these coordinates using algebraic manipulation becomes increasingly more difficult as the power increases. Hence, we can use our knowledge of differential calculus and apply it to solve such problems.

Before getting into any terminology, however, we must familiarise ourselves with the technical terms and associated definitions (see Figure 11 below).

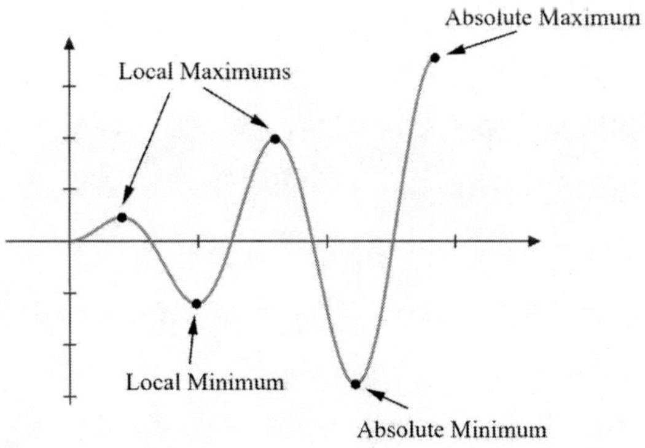

Fig. 11.

We are able to see that, at the second and fourth points, the (concave upwards) function reaches minimum points, while the (concave downwards) function reaches a maximum at the first, third, and fifth points. Concavity and its applications will be introduced in Part IV. It must be noted that, while these minimums and maximums exhibit similar properties, they remain fundamentally different, and more rigorous proofs using derivatives are needed to define them properly and solve for their unique cases.

Let us recall the Extreme Value Theorem, which lays the foundation for this topic.

RECALL

Let $f(x)$ be continuous over $[a, b]$;

$\exists$ maximum value of f and minimum value of f,
$\exists\, c, d \in [a, b]$: $f(c) \leq f(x) \leq f(d)$ for
$\forall x$ such that $x \in [a, b]$.

Continuing from the Extreme Value Theorem, let us define the maximum and minimum values, as well as the local (relative) and global (absolute) extrema.

KEY POINT

Let $f(x)$ be defined over $[a, b]$, such that $c \in [a, b]$:

A maximum, $f(c)$, is observed if $f(c) \geq f(x)$ for $\forall x, x \in [a, b]$;
A minimum, $f(c)$, is observed if $f(c) \leq f(x)$ for $\forall x, x \in [a, b]$.

KEY POINT

The maximum and minimum values are called extreme values, or extrema. The lowest and highest values of f over the interval are called absolute (global) extrema.

Local maximums can be identified when f' switches signs from + to −.

Local minimums can be identified when f' switches signs from − to +.

All absolute extrema are also relative extrema by definition*.

*Note: absolute extrema can occur at the endpoints of the interval; relative extrema can only occur within the open interval. Hence, if the highest/lowest value of the function is a critical point and at the endpoint, it can only be considered an absolute extremum, and not a local extremum. Furthermore, when both f' and f'' equal zero at a point, it is called a ***stationary (horizontal) point of inflection***, and not a relative extremum (see 4.2.). If the limit of the function is unbounded in the open interval on the positive and/or negative end, then the absolute extrema does not exist on that particular end. If both ends have unbounded limits (both positive and negative infinity), then the function only observes relative extrema.

This chapter will focus on relative extrema (local minimums and maximums) and how we can find them using our knowledge of differentiation.

EXAMPLE 3.3.1.

Find all local minimums of the function $f(x) = 4x^4 - 2x^2$.

SOLUTION

Let us analyse the critical points, c, of our function $f(x)$, which is defined over the entire domain.

$$f(x) = 4x^4 - 2x^2$$
$$f'(x) = 16x^3 - 4x$$

Substituting in 0 for f':

$$16x^3 - 4x = 0$$
$$x\left(16x^2 - 4\right) = 0$$
$$x = 0 \text{ or } 16x^2 - 4 = 0$$

$$4(4x^2 - 1) = 0$$
$$4x^2 - 1 = 0$$
$$4x^2 = 1$$
$$x^2 = \frac{1}{4}$$
$$x = \pm\frac{1}{2}$$

Therefore, there exists three critical points on f at $x = -\frac{1}{2}$, $x = 1$, and $x = \frac{1}{2}$.

A basic sign chart (without the interval calculations) can be shown as such:

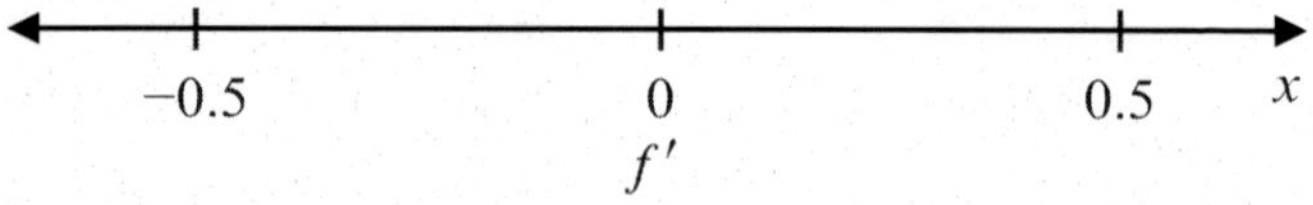

SOLUTION

Now, let us perform our interval calculations.For $\{-\infty < x < -0.5\}$: (substitute $x = -1$)

$$\begin{aligned} f'(x) &= 16x^3 - 4x \\ f'(-1) &= 16(-1)^3 - 4(-1) \\ &= -16 + 4 \\ &= -12\ (-ve) \end{aligned}$$

For $\{-0.5 < x < 0\}$: (substitute $x = -0.1$)

$$\begin{aligned} f'(-0.1) &= 16(-0.1)^3 - 4(-0.1) \\ &= -0.016 + 0.4 \\ &= 0.384\ (+ve) \end{aligned}$$

For $\{0 < x < 0.5\}$: (substitute $x = 0.1$)

$$\begin{aligned} f'(0.1) &= 16(-0.1)^3 - 4(0.1) \\ &= 0.016 - 0.4 \\ &= -0.384\ (-ve) \end{aligned}$$

For $\{0.5 < x < \infty\}$: (substitute $x = 1$)

$$\begin{aligned} f'(1) &= 16(1)^3 - 4(1) \\ &= 16 - 4 \\ &= 12\ (+ve) \end{aligned}$$

The following sign chart is derived from our calculations:

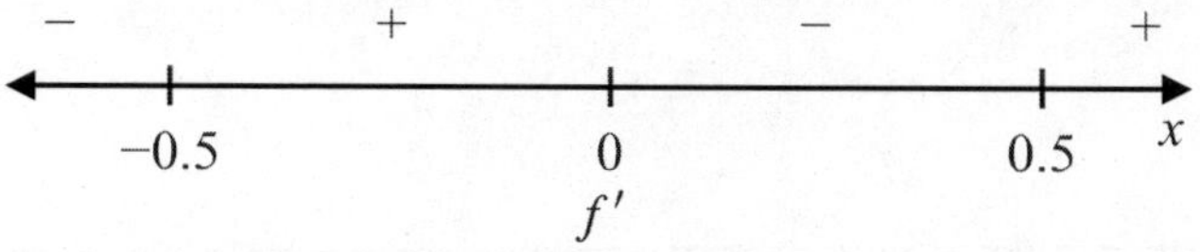

SOLUTION

There are two critical points where the derivative turns from negative to positive over the domain of f. Thus, the two local minimums of f exist at $x = -0.5$ and $x = 0.5$.

Let us try another example with some more peculiar functions.

EXAMPLE 3.3.2.

Find all local minimums of the function when $f'(x) = x^2(x+1)^2(x-1)^2$.

SOLUTION

The derivative is defined for all x in the domain, implying that f is continuous everywhere.

$$f'(x) = x^2(x+1)^2(x-1)^2$$

Substituting in 0 for f':

$$x^2(x+1)^2(x-1)^2 = 0$$
$$x^2 = 0 \text{ or } (x+1)^2 = 0 \text{ or } (x-1)^2 = 0$$

So:

$$x^2 = 0$$
$$x = 0$$

SOLUTION

Or:

$$\begin{aligned}(x+1)^2 &= 0\\ x+1 &= 0\\ x &= -1\end{aligned}$$

Or:

$$\begin{aligned}(x-1)^2 &= 0\\ x-1 &= 0\\ x &= 1\end{aligned}$$

$$x = -1 \text{ or } 0 \text{ or } 1$$

Therefore, there exist three critical points on f at $x = -1$, $x = 0$, and $x = 1$.

A basic sign chart (without the interval calculations) can be shown as such:

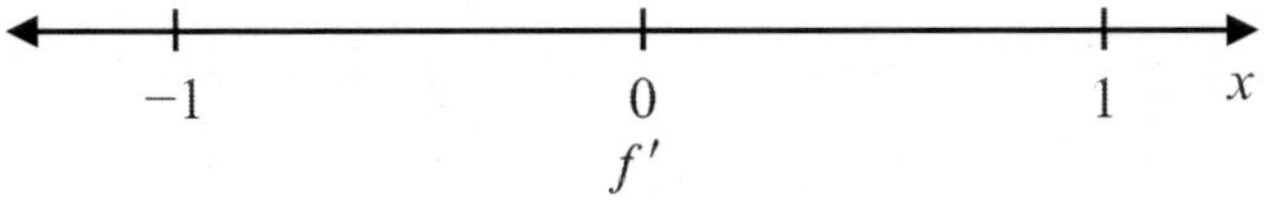

Now, let us perform our calculations to find the increasing and decreasing intervals of f.

For $\{-\infty < x < -1\}$: (substitute $x = -2$)

$$\begin{aligned}f'(x) &= x^2(x+1)^2(x-1)^2\\ f'(-2) &= (-2)^2(-2+1)^2(-2-1)^2\\ &= 4 \cdot 1 \cdot 9\\ &= 36\ (+ve)\end{aligned}$$

SOLUTION

For $\{-1 < x < 0\}$: (substitute $x = -0.5$)

$$f'(-0.5) = (-0.5)^2(-0.5+1)^2(-0.5-1)^2$$
$$= 0.25 \cdot 0.25 \cdot 2.25$$
$$= 0.140625\ (+ve)$$

For $\{0 < x < 1\}$: (substitute $x = 0.5$)

$$f'(0.5) = (0.5)^2(0.5+1)^2(0.5-1)^2$$
$$= 0.25 \cdot 2.25 \cdot$$
$$= 0.140625\ (+ve)$$

For $\{1 < x < \infty\}$: (substitute $x = 2$)

$$f'(2) = (2)^2(2+1)^2(2-1)^2$$
$$= 4 \cdot 9 \cdot 1$$
$$= 36\ (+ve)$$

The following sign chart is derived from our calculations:

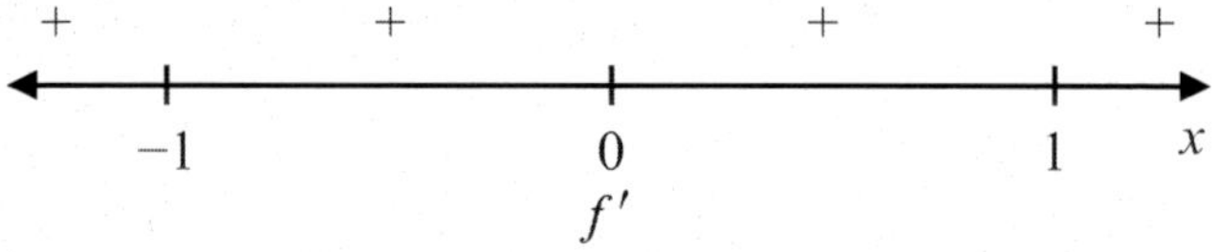

From the sign chart, we have deduced that f' is positive everywhere, meaning that the function is increasing for all $x \in \mathbb{R}$. Therefore, there are no extrema anywhere (note that absolute extrema do not exist either since the limits as $x \to \infty$ are unbounded), thus f exhibits no relative (local) minima.

3.3. Exercises

1. Find all extrema of the function $f(x) = x(4x+2)^2(x-4)^3$.

2. Find all local minima of $g(x) = -2\cos(x)$.

3. Find all local maxima of $h(x) = x^6 - 2x^5 + x^4$.

4. Find all extrema of $g(x) = 2x + x^2(x-2)(x+3)$.

5. Find all local maxima of $f(x) = -x^4(x-2)(x+5)$.

3.4. Absolute Extrema

Continuing from the previous chapter, absolute (global) extrema are the most positive and negative local extrema (on an open interval) or value (on the endpoints) that the function takes on within its domain. We can deduce absolute extrema by substituting in the x-values for local extrema and determining the maximum/minimum values, which are the global maximum and minimum respectively. This is known as the *Candidates Test*. Further analytical methods such as the *second derivative test* (see 4.3.1.) can be seen later in this book.

EXAMPLE 3.4.1.

Find the absolute extrema of the function $f(x) = 4x^4 - 4x^3 - 2x^2$.

SOLUTION

The function f is continuous across its domain, so we only need to worry about the derivative equalling zero. For our critical points:

$$f'(x) = 16x^3 - 12x^2 - 4x$$
$$0 = 16x^3 - 12x^2 - 4x$$

$$x\left(16x^2 - 12x - 4\right) = 0$$
$$x = 0 \text{ or } 16x^2 - 12x - 4 = 0$$

$$16x^2 - 12x - 4 = 0$$
$$4x^2 - 3x - 1 = 0$$
$$(x + 0.25)(4x - 4) = 0$$
$$x = -0.25 \text{ or } 1$$

SOLUTION

We have deduced:

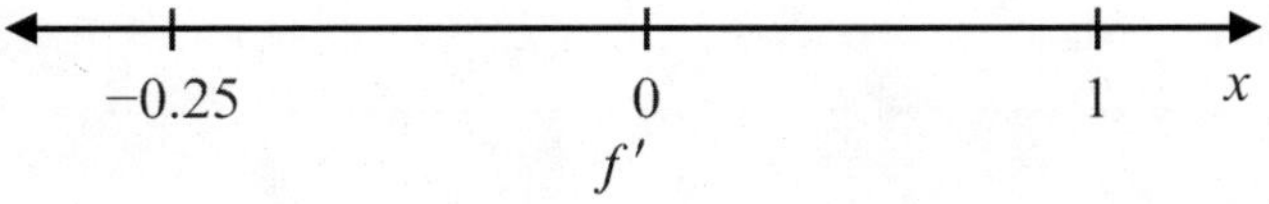

Substituting in values to find increasing/decreasing intervals:

For $\{-\infty < x < -0.25\}$: (substitute $x = -1$)

$$\begin{aligned} f'(x) &= 16x^3 - 12x^2 - 4x \\ f'(-1) &= 16(-1)^3 - 12(-1)^2 - 4(-1) \\ &= -16 - 12 + 4 \\ &= -24\ (-ve) \end{aligned}$$

For $\{-0.25 < x < 0\}$: (substitute $x = -0.1$)

$$\begin{aligned} f'(-0.1) &= 16(-0.1)^3 - 12(-0.1)^2 - 4(-0.1) \\ &= -0.016 - 0.12 + 0.4 \\ &= 0.264\ (+ve) \end{aligned}$$

For $\{0 < x < 1\}$: (substitute $x = 0.1$)

$$\begin{aligned} f'(0.1) &= 16(0.1)^3 - 12(0.1)^2 - 4(0.1) \\ &= 0.016 - 0.12 - 0.4 \\ &= -0.504\ (-ve) \end{aligned}$$

For $\{1 < x < \infty\}$: (substitute $x = 2$)

$$\begin{aligned} f'(2) &= 16(2)^3 - 12(2)^2 - 4(2) \\ &= 128 - 48 - 8 \\ &= 72\ (+ve) \end{aligned}$$

SOLUTION

Thus, the following sign chart was constructed:

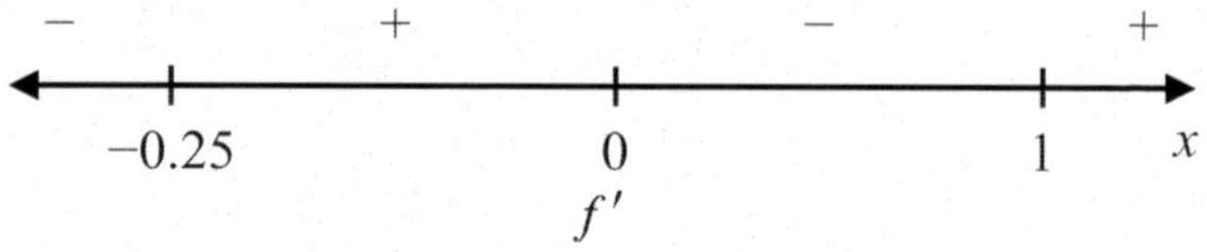

We can observe two local minimums of f exist at $x = \{-0.25, 1\}$, and a local maximum at $x = 0$.

To find the absolute minimum, we can simply substitute the x-value of critical point into f:

$$f(x) = 4x^4 - 4x^3 - 2x^2$$

$$\begin{aligned} f(-0.25) &= 4(-0.25)^4 - 4(-0.25)^3 - 2(-0.25)^2 \\ &= -0.047 \end{aligned}$$

$$\begin{aligned} f(1) &= 4(1)^4 - 4(1)^3 - 2(1)^2 \\ &= -2 \end{aligned}$$

As -2 is less than -0.047, the absolute minimum resides at $x = 1$. Therefore, the absolute minimum of f exists at $x = 1$. The absolute maximum does not exist as the limits of the function which tend toward $\pm\infty$ are unbounded.

Let us try an example where the function is undefined at some point in the domain.

EXAMPLE 3.4.2.

Find the absolute extrema of the function $h(x) = x^2\ln(x)$.

SOLUTION

The natural logarithm only returns a value if x is a positive real number. Hence the domain is limited to $x > 0$. Let us find the derivative using the product rule:

$$\begin{aligned} h(x) &= x^2 \ln(x) \\ h'(x) &= u'v + v'u \\ &= 2x\ln(x) + x \\ &= x(2\ln(x) + 1) \end{aligned}$$

Hence, from the equation, we have deduced that:

$$x = 0 \text{ or } 2\ln(x) + 1 = 0$$

Note: x cannot equal 0 as implied by the domain.

$$\begin{aligned} 2\ln(x) + 1 &= 0 \\ 2\ln(x) &= -1 \\ \ln(x) &= -0.5 \\ x &= e^{-0.5} \\ &\approx 0.607 \end{aligned}$$

We have deduced:

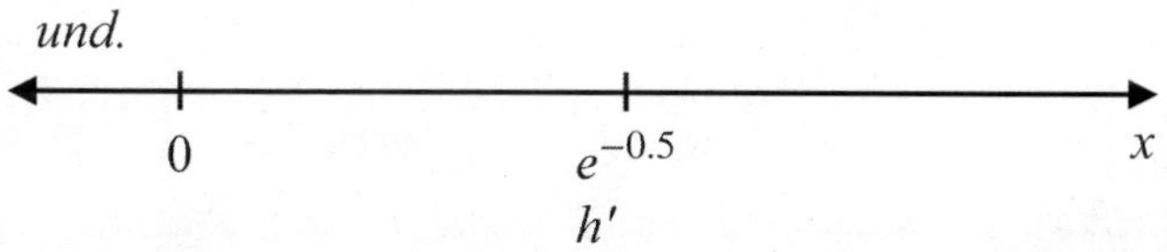

SOLUTION

Substituting in values to find the intervals:

For $\{0 < x < e^{-0.5}\}$: (substitute $x = 0.1$)

$$\begin{aligned} h'(x) &= 2x\ln(x) + x \\ h'(0.1) &= 2(0.1)\ln(0.1) + 0.1 \\ &= -0.361\ (-ve) \end{aligned}$$

For $\{e^{-0.5} < x < \infty\}$: (substitute $x = 1$)

$$\begin{aligned} h'(1) &= 2(1)\ln(1) + 1 \\ &= 1\ (+ve) \end{aligned}$$

Thus, the following sign chart is derived:

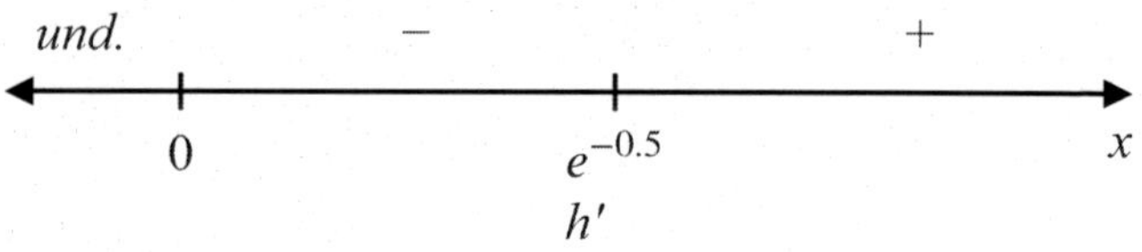

Only one local minimum exists on h before it increases perpetually, observable at $e^{-0.5}$. Thus, it is the absolute minimum. The limit of h as $x \to \infty$ is unbounded in the positive direction, meaning no absolute (and no relative) maximum exists for h.

3.4. Exercises

1. Find the absolute extrema of the function $f(x) = x(x - 1)(x + 1)(2x + 2)$.

2. Daniel is aspiring to compete in the 2032 Brisbane Olympic Games for badminton. One day, he played in a very windy hall, where the shuttlecock's motion on a high serve can be modelled by the equation $S(x) = -4(x - 1)^4 + 4(x - 1)^2 + x + 1.1$, assuming both the x and y axes are measured in metres.
   a) What is an appropriate domain for the equation?
   b) Find the instances when the shuttlecock only had movement in the horizontal (positive x) direction.
   c) Find the highest the shuttlecock travelled during the serve.
   d) If the shuttlecock had an initial velocity (in both directions) of 9ms^{-1}, how long did it take for the shuttlecock to reach the maximum height after it was struck? (Hint: for the angle at which the shuttlecock travelled after the serve, find the slope at $x = 0$, then use the inverse tangent function.)

3. Find the absolute extrema of the function $y = x^4 - 2x^3 - 2x^2 + x, x \in [-1, 1]$.

4. **Extension**: Find all critical points and the absolute extrema of the function $g(x) = \frac{3x}{x + 4} - \frac{x^3}{x - 2} - \frac{x}{x - 5} - 2x + 4$.

PART IV

Concavity, Differentials, and Antidifferentiation

4.1. Introduction to Concavity

Concavity describes the phenomena of second derivatives associated with the visual portions of a graph. It involves the application of techniques such as sign charts (seen in Part III), and the second derivative test to find intervals of concavity and other related properties of functions.

Let us recall the definition of a second derivative.

RECALL

$$\frac{d}{dx}\frac{dy}{dx} = \frac{d^2y}{dx^2} = y''$$

In a more understandable context, we can analyse the rate of change of the function's first derivative – this is known as concavity.

If a function's second derivative is positive, we call the interval 'concave upwards'; if a function's second derivative is negative, then it is 'concave downwards'. This also applies for open intervals.

KEY POINT

Let $f(x)$ be defined over $[a, b]$:

$f''(x) < 0 \Longrightarrow$ concave downwards (∩-shaped);
$f''(x) > 0 \Longrightarrow$ concave upwards (∪-shaped).

Concavity can only change if the second derivative changes signs.

Let us take a look at how $f(x)$ behaves in Figure 12 below.

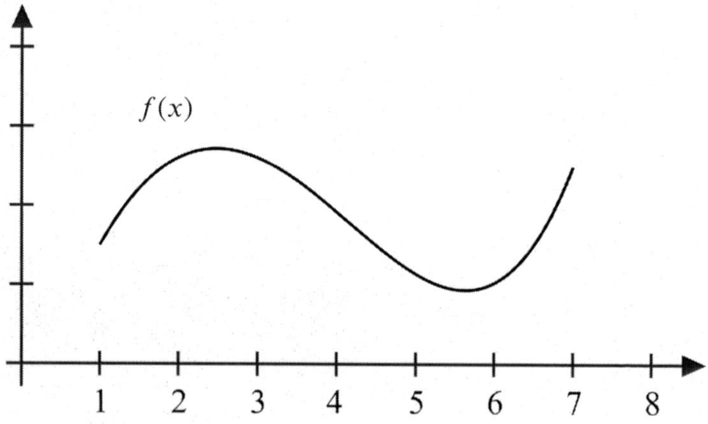

Fig. 12.

It is assumed that the function $f(x)$ is at least twice differentiable with respect to x. The function is defined and continuous over $x \in [1, 7]$. We can obviously see our absolute maximum and minimum of f, between $x = \{2, 3\}$ and $x = \{5, 6\}$ respectively. However, when we talk about the behaviour of f, we also want to note the intervals of concavity and when it switches.

We can conclude from inspection that f appears to be concave downwards over $x \in [1, 4)$, and concave upwards between $x \in (4, 7]$. Note that we do not include $x = 4$ in the set notation because it is a *point of inflection*. This means that $f''(x) < 0$ when $x \in [1, 4)$, and $f''(x) > 0$ for $x \in (4, 7]$. In other words, f' becomes increasingly negative until the inflection point, where it begins to increase (but still remains negative). The second derivative changes signs over the inflection point at $x = 4$, turning positive.

There is another definition for concavity, and it has to do with the tangent line on the graph. If the tangent line lies below the function, it is concave upwards, and vice versa. Figure 13 below visualises this.

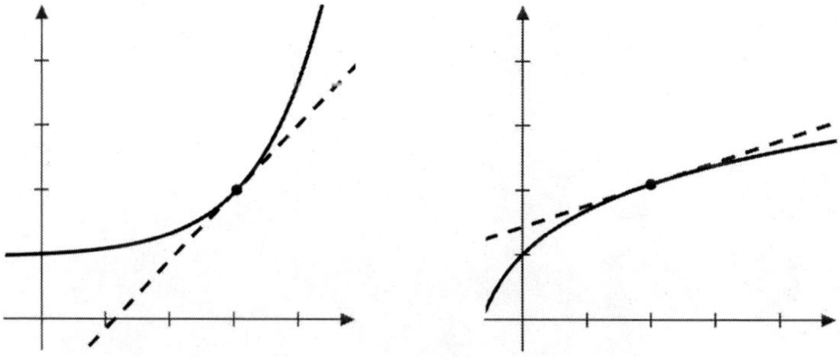

a. *b.*

Fig. 13.

4.1. Exercises

1. Deduce when the function $f(x) = \frac{1}{2}x^3$ is concave upwards, graphically.

2. Deduce when the function $g(x) = -\sin(x), x \in (0, 2\pi)$ is concave downwards, graphically.

3. Deduce when the function $f(x) = x^2(x^2 + 1)$ is concave upwards and concave downwards, graphically.

4.2. Inflection Points

A point of inflection, also known as an inflection point, is a point on the graph where the second derivative changes signs. The process of calculating inflection points is similar to calculating our critical points, only applied to the second derivative. Aforementioned, the function's concavity (curvature) can change over an inflection point; in order for the second derivative to change signs, the point must have $f''(x) = 0$ or $f''(x)$ undefined. However, the inflection point itself must be defined. Let us denote an inflection point as $(c, f(c))$ on a function f.

KEY POINT

Let $f(x)$ be defined over $[a, b]$, such that $c \in [a, b]$:

An inflection point is viewed at $x = c$ if both of the following criteria are met:

$f''(c) = 0$, and $f''(x)$ changes from + to – or – to + across $f''(c)$*.

A stationary (horizontal) point of inflection is present if both $f'(c) = 0$ and $f''(c) = 0$.

*Note: if $f''(c) = 0$, but the second derivative does not change signs, it is not an inflection point, but rather an undulation point. Mathematically, an inflection point also requires the next highest non-zero derivative to be of odd order (i.e. $f'''(c), f'''''(x) \neq 0$). If the fourth, sixth, or eighth derivative is non-zero, where the fifth, seventh, and ninth derivatives are zero, then it is called a point of undulation. However, most of the questions in this book will focus on the simpler definitions, and knowledge of undulation points is not required. If the second derivative is undefined at a point, but the curvature changes around it, then the inflection point does not exist, but concavity switches nonetheless.

Let us try an example below.

EXAMPLE 4.2.1.

Analyse the concavity and find all inflection points of the function $f(x) = 2x^2 - 3x^3$.

SOLUTION

Similar to the process of finding critical points, we start out by finding the second derivative of our function f:

$$\begin{aligned} f''(x) &= \frac{d}{dx}\frac{d}{dx}\left[2x^2 - 3x^3\right] \\ &= \frac{d}{dx}\left[4x - 9x^2\right] \\ &= 4 - 18x \end{aligned}$$

Now, let us substitute $f'' = 0$ into our equation:

$$\begin{aligned} 4 - 18x &= 0 \\ 18x &= 4 \\ x &= \frac{4}{18} \\ &= \frac{2}{9} \\ &\approx 0.22 \end{aligned}$$

We can now create a sign chart for our second derivative calculation.

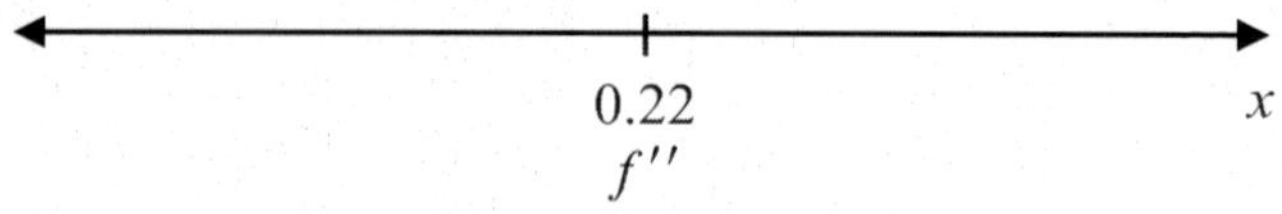

SOLUTION

Now, we can substitute in values of x to analyse the function's concavity:

For $\{-\infty < x < 0.22\}$: (substitute $x = 0$)

$$\begin{aligned} f''(0) &= 4 - 18x \\ &= 4 - 18(0) \\ &= 4\ (+ve) \end{aligned}$$

For $\{0.22 < x < \infty\}$: (substitute $x = 1$)

$$\begin{aligned} f''(1) &= 4 - 18x \\ &= 4 - 18(1) \\ &= -14\ (-ve) \end{aligned}$$

Thus, the following sign chart is derived:

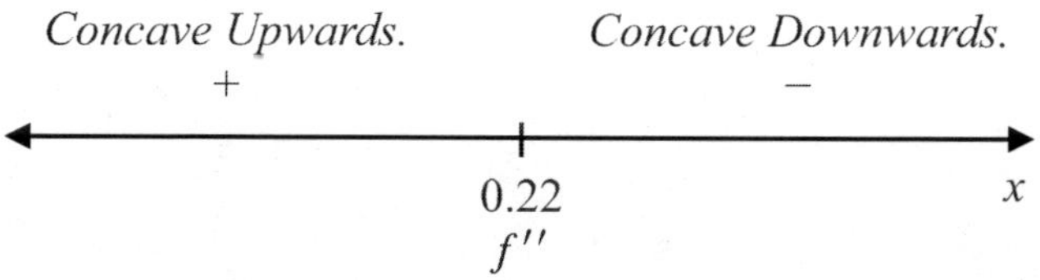

Therefore, the interval $x \in \left(-\infty, \frac{2}{9}\right)$ is concave upwards, while the interval $x \in \left(\frac{2}{9}, \infty\right)$ is concave downwards, with a single inflection point at $x = \frac{2}{9}$.

4.2. Exercises

1. Analyse the concavity and find the inflection points of the function $g(x) = -x^4 + 6x^2 - 2x$.

2. Analyse the concavity and find the inflection points of the function $f(x) = \frac{1}{4}x^4 - 4x^3 + 24x^2$.

3. Analyse the concavity and find the inflection points of the function $f(x) = -\frac{1}{x+3}$.

4. **Extension**: Analyse the concavity and find the inflection points of the function $f(x) = x^4 - x$.

4.3. Optimisation

Optimisation is very important in real-life applications because it can minimise errors and costs or determine the best parameters for theoretical modelling. To do so, we will apply our knowledge of critical points, relative extrema, and concavity to reach the best possible solution, usually by maximising or minimising our variables.

4.3.1. Second Derivative Test

The second derivative test states that, for critical points where $f'(c) = 0$, where f' is defined around $x = c$ and $f''(c)$ exists, the concavity of the function will determine whether $(c, f(c))$ is a local minimum or maximum. While this seems intuitive, given the previous knowledge we have acquired of second derivatives, there are still many nuances to this topic which we will uncover.

KEY POINT

Given a twice differentiable function where $f'(c) = 0$ and f' exists in the neighbourhood of $x = c$ such that $f''(c) \in \mathbb{R}$:

$$f''(c) < 0 \implies \text{relative maximum at } x = c,$$
$$f''(c) > 0 \implies \text{relative minimum at } x = c,$$

$$f''(c) = 0 \implies \text{inconclusive (more information required).}$$

(Note: the first derivative test is always conclusive, but the second derivative test may be inconclusive at times.)

A typical exam question may ask you to determine whether a point is a relative minimum or maximum using this test, and we will try a similar format question below.

EXAMPLE 4.3.1.1.

Let $g(x)$ be a twice differentiable function, where $g(2) = 2$, $g'(2) = 0$, and $g''(2) = -3$. What does the function exhibit at $(2, 2)$?

SOLUTION

Since $g'(2) = 0$, we know that the point $(2, 2)$ is a critical point. The question also gives the second derivative, $g''(2)$, which is equal to -3.

We know that a negative second derivative means the interval is concave downwards. Intuitively, we know that the concave downwards interval has a peak, where the function takes upon the greatest value in its neighbourhood. Thus, we know that a negative second derivative implies a local maximum*.

Therefore, the point $(2, 2)$ is a relative/local maximum on the function g.

**Note: the reason why we used intuition first to conclude the answer (instead of simply noting the key point explanation above) is because fully understanding the concept is much more useful than rote learning facts. Especially in differential calculus, a topic with very deep and intricate subject matter, we should attempt to use these knowledge points to further our understanding instead of merely remembering key points for the sake of memorisation.*

4.3.2. Function Optimisation

When we discuss function optimisation, we are usually given two variables, x and y, and a set of parameters for us to manipulate into a primary and a secondary equation. In most cases, we want to find (absolute) minimums and maximums, which can be achieved through evaluating the critical points and second derivatives.

There are many types of optimisation problems, including finding maximum profit, minimum surface area, as well as some Olympiad-type questions related to the processing of numbers. Let us try a simple 'sum-of-squares' problem below.

EXAMPLE 4.3.2.1.

Mr. Coyne has x apples and y oranges. Ms. Gleeson makes the observation that the number of apples multiplied by the number of oranges is equal to 36. Find the lowest possible value of $x^2 + y^2$.

SOLUTION

After interpreting the question, we need to construct a primary and secondary set of equations that best represents the given information. The secondary equation eliminates one variable by representing it as a factor of the other, which we can use to substitute into the primary equation later on. For example:

$$xy = 36$$

$$y = \frac{36}{x}$$

SOLUTION

Now, to construct the primary equation, we substitute the secondary equation into the y variable and make it a function.

Let us denote the sum function as $S(x)$.

$$\begin{aligned} S(x) &= x^2 + y^2 \\ &= x^2 + \left(\frac{36}{x}\right)^2 \\ &= x^2 + 1296x^{-2} \end{aligned}$$

Now that we have our appropriate function, we can apply our calculus knowledge to find the lowest possible for S, which is just an absolute minimum calculation.

Let us calculate for $S'(x)$:

$$\begin{aligned} S(x) &= x^2 + 1296x^{-2} \\ S'(x) &= 2x - 2592x^{-3} \\ 0 &= 2x - 2592x^{-3} \\ 2x &= 2592x^{-3} \\ 2x^4 &= 2592 \\ x^4 &= 1296 \\ x &= \pm 6 \end{aligned}$$

Since we are dealing with whole numbers of apples and oranges, the value of x must be a natural number, so there is a single useful critical point at $x = 6$.

SOLUTION

Now, to confirm whether S has an absolute minimum at $x = 6$, we want to check its concavity using the second derivative test.

$$\begin{aligned} S''(x) &= \frac{d}{dx}\left[2x - 2592x^{-3}\right] \\ &= 2 + 7776x^{-4} \\ S''(6) &= 2 + 7776(6)^{-4} \\ &= 2 + 6 \\ &= 8\ (+ve) \end{aligned}$$

We have a positive second derivative at $x = 6$, meaning it is concave upwards, and hence a minimum. We know it is an absolute minimum because no other critical points are present over the useful domain of $x \in \mathbb{N}$.

Now, let us answer the proposed question by evaluating y and $x^2 + y^2$.

$$\begin{aligned} y &= \frac{36}{x} \\ &= \frac{36}{6} \\ &= 6 \end{aligned}$$

$$\begin{aligned} x^2 + y^2 &= 6^2 + 6^2 \\ &= 36 + 36 \\ &= 72 \end{aligned}$$

Therefore, the minimum value of $x^2 + y^2$ when $xy = 36$ is equal to 72.

Let us try a question related to the volume of a shape and its respective surface area.

EXAMPLE 4.3.2.2.

Ethan loves drinking soft drinks. His favourite soft drink is contained within a large can of radius 5cm and height 12cm. He wondered if changing the dimensions, but retaining the surface area, could alter the practicality and usefulness of the can in question.

a) Find the maximum volume that a can of the same surface area can hold, including its dimensions. Assume the can is a perfect, closed cylinder.
b) Would this be practical in real-life applications; what is its suitability compared to Ethan's favourite can?

SOLUTION

Firstly, we need to find the surface area of the can in question:

$$\begin{aligned} S.A. &= 2\pi r^2 + 2\pi rh \\ &= 2\pi(5)^2 + 2\pi(5)(12) \\ &= 50\pi + 120\pi \\ &= 170\pi \text{cm}^2 \end{aligned}$$

Now that we have our surface area, we can construct our secondary equation.

SOLUTION

For the secondary equation, we want to express one variable in terms of the other. Since radius is the more important constraint, we will notate height in terms of the radius.

$$\begin{aligned}170\pi &= 2\pi r^2 + 2\pi rh \\ 170\pi &= 2\pi\left(r^2 + rh\right) \\ 85 &= r^2 + rh \\ rh &= 85 - r^2 \\ h &= \frac{85 - r^2}{r}\end{aligned}$$

This is our derived secondary equation. Now, we can substitute the above equation into our volume formula, making it a function of the radius.

$$\begin{aligned}V &= \pi r^2 h \\ V(r) &= \pi r^2\left(\frac{85 - r^2}{r}\right) \\ V(r) &= \pi r\left(85 - r^2\right) \\ V(r) &= \pi(85r - r^3)\end{aligned}$$

To find our critical points:

$$\begin{aligned}V'(r) &= \pi \cdot \frac{d}{dr}\left[85r - r^3\right] \\ &= \pi\left(85 - 3r^2\right) \\ 0 &= \pi\left(85 - 3r^2\right) \\ 0 &= 85 - 3r^2 \\ 3r^2 &= 85 \\ r^2 &= \frac{85}{3} \\ r &\approx \pm 5.32\text{cm}\end{aligned}$$

SOLUTION

Again, we are only dealing with positive values; hence, the optimal radius of the can is approximately 5.32cm.

Let us confirm this using the second derivative test:

$$\begin{aligned} V''(r) &= \frac{d}{dr}\left[\pi\left(85 - 3r^2\right)\right] \\ &= \pi \cdot -6r \\ &= -6\pi r \\ V''(5.32) &= -6\pi \cdot 5.32 \\ &= -100.33\ (-ve) \end{aligned}$$

Since the second derivative is negative, we know that it is an absolute maximum in this case.

To answer part a), we need to find the height of the can and the volume.

$$\begin{aligned} h &= \frac{85 - r^2}{r} \\ &= \frac{85 - (5.32)^2}{5.32} \\ &\approx 10.65\text{cm} \end{aligned}$$

$$\begin{aligned} V &= \pi r^2 h \\ &= \pi \cdot 5.32^2 \cdot 10.65 \\ &\approx 947.60\text{cm}^3 \\ &\approx 947.60\text{mL} \end{aligned}$$

Therefore, the can has a maximum volume of 947.60mL, with a radius of 5.32cm and a height of 10.65cm.

SOLUTION

To answer part b): this model is still relatively practical in a real-life context and could possibly be adopted for a more efficient vessel design. However, limitations arise in the assumption that the can is a perfect cylinder.

4.3. Exercises

1. For an arbitrary, twice differentiable function $P(x)$, where $x \in [-1, 4]$, $P(1) = 0$, $P'(1) = 0$, and $P''(1) = 2$. What is observed at the point $(1, 0)$?

2. For an arbitrary, twice differentiable function $f(x)$, where $x \in \mathbb{R}$, $f(2) = 5$, $f'(2) = 0$, and $f''(2) = 0$. What is observed at the point $(2, 5)$?

3. James owns a fishing business, and his profit is modelled by the time function $P(t) = -\frac{t}{800}(t - 50)(t^2 - 10t)$ (time is measured in hours).
   a) At what time does he make the greatest profit?
   b) At what time does he lose the most money?

4. **Extension**: A rectangular cardboard sheet, 16cm by 12cm, has four squares of side length x cm cut out from its four corners to create a net for an open-lid box.
   a) Derive a general equation for the box's volume, $V(x)$.
   b) What is an appropriate domain for the equation?
   c) Find the maximum possible box volume.
   d) Find the minimum possible box volume.

4.4. Implicit Relations

When we refer to implicit relations in differential calculus, we really mean the applications of techniques to solve questions requiring implicit differentiation rather than just the latter. This may include recognising certain properties of graphs and/or finding horizontal/vertical tangent lines to the implicit relation.

For introductory purposes, we will mainly focus on the tangent lines. Let us try an example problem.

EXAMPLE 4.4.1.

An implicit relation has the equation $x^2 + y^4 - 10x = 11$.

a) Find where the relation has an instantaneous slope of 0.
b) Find the points where the relation has a vertical tangent line

SOLUTION

Firstly, let us implicitly differentiate the equation:

$$x^2 + y^4 - 10x = 11$$
$$\frac{d}{dx}\left[x^2 + y^4 - 10x\right] = \frac{d}{dx}[11]$$
$$2x + 4y^3\frac{dy}{dx} - 10 = 0$$
$$4y^3\frac{dy}{dx} = 10 - 2x$$
$$\frac{dy}{dx} = \frac{10 - 2x}{4y^3}$$

Therefore, the derivative of the implicit relation with respect to x is $\frac{10-2x}{4y^3}$.

SOLUTION

To answer part a), we need to find when the derivative equals 0:

$$\begin{aligned}\frac{10-2x}{4y^3} &= 0\\ 10-2x &= 0\\ 2x &= 10\\ x &= 5\end{aligned}$$

Therefore, the line tangent to the relation is horizontal at $x = 5$. For b), we need to find when the derivative is undefined:

$$\begin{aligned}\frac{10-2x}{4y^3} &= \text{undefined}\\ 4y^3 &= 0\\ y &= 0\end{aligned}$$

Substituting back into the relation:

$$\begin{aligned}x^2+y^4-10x &= 11\\ x^2+0^4-10x &= 11\\ x^2-10x &= 11\\ x^2-10x-11 &= 0\end{aligned}$$

$$\begin{aligned}x^2-10x-11 &= 0\\ (x-11)(x+1) &= 0\\ x &= -1 \text{ or } 11\end{aligned}$$

The derivative is undefined at $x = -1$ or $x = 11$. Therefore, the points at which the derivative is undefined are $(-1, 0)$ and $(11, 0)$.

4.4. Exercises

1. An implicit relation is given by the equation $(x-6)^4 + (y+2)^6 = 1$.
   a) Find where the relation has an instantaneous slope of 0.
   b) Find the points where the relation has a vertical tangent line.

2. A circle has the equation $(y-7)^2 + (x+1)^2 = 16$.
   a) Find where the relation has an instantaneous slope of 0.
   b) Find the points where the relation has a vertical tangent line.

3. An implicit relation has the equation $x^2 + xy + y^2 = 9$. Find the absolute extrema.

4. **Extension**: An implicit relation has the equation $y^2 = x^2y^3$. Find when the tangent line is parallel to the line $y = -x$.

5. **Extension**: Implicitly differentiate $4y^6 - y^3 + 2x^2y^2 + 14xy = 3x^5y^2 - 12y^9$ with respect to x.

4.5. Differentials and Integral Calculus

We have mentioned the notion of integration previously in this book, when referring to the relationship between various vector quantities in physics, although it has not been properly quantified nor explained. It is fitting however, at the end of a differentiation textbook, that we touch upon the notion of integral calculus and its applications. Simply, integration is the process that finds the area under the curve, or between two functions, and is the reverse of differentiation.

Before we discuss integration in more detail, we want to understand differentials and their relationship to integral calculus, to avoid future confusion of these topics.

4.5.1. Differentials

In this section, we will be familiarising ourselves with some useful notation and its applications.

Let $y = f(x)$, and $\frac{dy}{dx} = f'(x)$. We can rewrite the latter as:

$$\boxed{dy = f'(x)\, dx}$$

Note: dy and dx are what we defined as differentials at the beginning; for an arbitrary function $\phi(x)$, we want to compute the differential $d\phi$ (and vice versa for other functions).

EXAMPLE 4.5.1.1.

Compute the differential for the following functions:

a) $f(x) = 3x^2 + 5x - 2$

b) $y(t) = t^3 - 3t^2 + 5t$

SOLUTION

For Question a):

$$f(x) = 3x^2 + 5x - 2$$
$$f'(x) = 6x + 5$$

$$dy = f'(x)\,dx$$
$$dy = (6x + 5)dx$$

For Question b):

$$y(t) = t^3 - 3t^2 + 5t$$
$$y'(t) = 3t^2 - 6t + 5$$

$$dy = y'(t)\,dt$$
$$dy = \left(3t^2 - 6t + 5\right)dt$$

Now that we have learnt how to compute a function's differential, let us explore some interesting properties and applications of the concept.

EXAMPLE 4.5.1.2.

Calculate Δy and dy for the function $y = x^2 + 2x + 4$ as x increases from 1 to 1.02.

SOLUTION

To demonstrate this property, we need to recall $\Delta y = f(x + \Delta x) - f(x)$. You may notice that this definition is eerily similar to the limit definition of a derivative – this is because we can regard $\Delta y \approx dy$ as $\Delta x \to 0$, as demonstrated by $\frac{dy}{dx} = \frac{f(x+h) - f(x)}{h}$, where $h = \Delta x$.

In this context, $dx = 1.02 - 1 = 0.02$.

Firstly, let us compute the value for Δy:

$$\begin{aligned} \Delta y &= f(x + \Delta x) - f(x) \\ &= (1.02)^2 + 2(1.02) + 4 - \left[(1)^2 + 2(1) + 4\right] \\ &= 7.0804 - 7 \\ &= 0.0804 \end{aligned}$$

Now, let us find dy:

$$\begin{aligned} dy &= y'(x)\, dx \\ &= (2x + 2)dx \\ &= (2(1) + 2)(0.02) \\ &= 0.08 \end{aligned}$$

Comparing Δy and dy, it is apparent that their values are very close. In theory, both values will converge to 0 as $\Delta x \to 0$, which both Δy and dy will take upon infinitesimally small values, and thus we arrive at our first principles yet again.

Let us put this application of differentials into practice with a standard percentage error question on the next page.

EXAMPLE 4.5.1.3.

A cube of length 5cm was measured with a possible uncertainty of ± 0.01cm. Find the maximum percentage error in the volume if we assume the length is exactly 5cm?

SOLUTION

Let us first differentiate the volume of a cube:

$$\begin{aligned} V(x) &= x^3 \\ V'(x) &= 3x^2 \end{aligned}$$

Now, we get the differential:

$$dV = 3x^2 dx$$

Assuming $\Delta y \approx dy$, we can compute dV:

$$\begin{aligned} dV &= 3(5)^2(0.01) \\ &= 0.75\text{cm}^3 \end{aligned}$$

To find percentage error:

$$\frac{0.75}{5^3} \cdot 100\% = 0.6\%$$

Therefore, there is a maximum percentage error (for volume) of 0.6% when taking 5cm to be the side length.

While the applications of differentials are useful in certain cases (such as in the example above), the main focus is to concentrate our knowledge on integration, which is much more multifaceted.

4.5.2. Antidifferentiation

Let us recall the definition of a computed differential for $f(x)$.

RECALL

$$df = f'(x)\,dx$$

If we wished to return the above differential to its original form (expressed in the form of $f(x)$), we would integrate both sides to get the following ($\int$ is the integral sign):

$$\int df = \int f'(x)\,dx$$

$$\boxed{f(x) = \int f'(x)\,dx}$$

This is a primitive definition of the integration process and what we mean when we say the derivative and the integral are inverses of each other – the integral sign 'cancels out' the differential. The main takeaway from differentials is how we arrived at the line below.

KEY POINT

Given a differentiable function $f(x)$, the *antiderivative* of $f(x)$ is any function $F(x)$ where:

$$F'(x) = f(x)$$

The general case for the antiderivative of $f(x)$ is called an *indefinite integral.*

KEY POINT

An indefinite integral is denoted as:

$$\int f(x)\,dx = F(x) + C$$

Where:
$f(x)$ is the original function (called the integrand),
$F(x)$ is the indefinite integral of $f(x)$, and
C * is an arbitrary constant (called the constant of integration).

*Note: the reason why we must add the constant of integration onto the end of an indefinite integral is because the derivative of a constant disappears during differentiation.

Now, with some intuition, we can compute some basic indefinite integrals.

EXAMPLE 4.5.2.1.

Evaluate the indefinite integrals for the following:

a) $\int 2x + 2\,dx$
b) $\int x^3 + 4x + 3\,dx$

SOLUTION

For Question a):

Before we learn any methods of integration, we can use intuition to integrate some basic polynomials.

SOLUTION

Considering that $2x + 2$ is what we integrate, we want to consider what function did we differentiate to get to $2x + 2$.

We know that $\frac{d}{dx}x^2 = 2x$ and $\frac{d}{dx}2x = 2$. Therefore:

$$\int 2x + 2\,dx$$
$$= x^2 + 2x + C$$

It is important to note the location of the differential dx, as $\int 2x + 2\,dx \neq \int 2x\,dx + 2$. Wherever we drop the dx is where we wish to stop integrating the function. Think of the integral sign and the differential as a pair of parentheses, where $\int$ denotes an open bracket, and the dx closes it.

For Question b):

$\frac{d}{dx}\frac{1}{4}x^4 = x^3$, $\frac{d}{dx}2x^2 = 4x$, $\frac{d}{dx}3x = 3$. Therefore:

$$\int x^3 + 4x + 3\,dx$$
$$= \frac{1}{4}x^4 + 2x^2 + 3x + C$$

Before the end of this book, it is appropriate that we learn the basic rules of integration, as we did with derivatives back in Part I.

The most fundamental processes are the reverse power rule and the constant rule, and are exactly what the names suggest, just the inverse operation of the derivative.

KEY POINT

The reverse power rule states:

$$\int x^n \, dx = \frac{x^{n+1}}{n+1} + C, n \neq -1$$

The constant rule states:

$$\int k \, dx = kx + C, k \text{ is a constant.}$$

There are also some noteworthy properties of integrals, similar to the derivatives:

KEY POINT

The addition and subtraction properties of integrals states:

$$\int f(x) \pm g(x) \, dx = \int f(x) \, dx \pm \int g(x) \, dx$$

The constant multiplication property states:

$$\int k f(x) \, dx = k \cdot \int f(x) \, dx$$

However, the product and quotients of functions are not the integrals multiplied and divided respectively. They have separate rules (just like the product and quotient rules of derivatives).

KEY POINT

$$\int f(x)g(x) \neq \int f(x)\,dx \cdot \int g(x)\,dx$$

$$\int \frac{f(x)}{g(x)} \neq \frac{\int f(x)\,dx}{\int g(x)\,dx}$$

These rules will not be covered in this book.

4.5. Exercises

1. Compute the differential for $f(x) = 4\ln(x) + x^2$.

2. Compute the differential for $h(t) = 15t^3 + 12t^2 - 3t^{-1} + 2$.

3. Find Δy and dy for $y = 3x^4 + 2x^2 + 2$ as x increases from 0.99 to 1.

4. Find Δf and df for $f(x) = 2\cos(x + 2) - x$ as x increases from 0 to 0.01.

5. Prove that $\lim_{\Delta x \to 0} \Delta y = dy$.

6. Evaluate $\int x^4 + 4x + 7\,dx$.

7. Evaluate $\int 2x^3\,dx + \int 3x\,dx$.

8. Evaluate $\int 2x\,dt$.

9. **Extension**: Research definite integrals. Evaluate $\int_{10}^{13} 2x\,dx$.

10. **Extension**: An object travels in a line. Between 0 and 5 seconds, it accelerates uniformly to $20ms^{-1}$ from rest. Then, it remains at the constant velocity for 10 seconds. Finally, it decelerates at a constant rate back to rest for 10 seconds. Find the general expression for the definite integral that describes the object's displacement.

11. **Extension**: Find the final displacement for the object in Question 10.

12. **Extension**: Rewrite $a(t) = v'(t) = s''(t)$ in integral notation. Assume no initial velocity.

Notes on the Exercises

PART I

CHAPTER 1.3.

1.

a) 3
b) 5
c) 10
d) 10
e) 1
f) 0

2.

a) $2x$
b) $9x^2$
c) $-3x^2$
d) $-9x^2 + 24x + 1$
e) $-4x + 4$
f) $5x^4 + 4x^3 + 3x^2$

3.

a) $-x^2$
b) $-9x^{-4} + 4x^{-3}$
c) $3x^{-4} + 2x$
d) $15x^{-4} - x^{-2} + 4x$

4.

No. A function can be continuous but not differentiable at a point (e.g. sharp point).

5.

We cannot assign x a value as it is a variable, not a constant.

CHAPTER 1.4.

1.

a) $60x$ b) $420x^6$

c) 20 d) $250x^4$

2.

a) $-15x^{-4}$ b) $24x^{11} + 1.5x^{-2}$

c) $-(x^{-5})(6x + 20)$ d) $-\frac{14}{x^5}$

3.

See 1.6. The Chain Rule.

CHAPTER 1.5.

1.

a) $\sin(x) + x\cos(x)$

b) $x(2\cos(x) - x\sin(x))$

c) $x^{-2}\left(x\sin(x) - \cos(x)\right)$

d) $\sec(x)(\tan^2(x) + \sec^2(x))$

e) $\csc(x)(\cot^2(x) + \csc^2(x))$

f) $2\sec^2(x)$

2.

a) $\frac{x\cos(x) - \sin(x)}{x^2}$

b) $\frac{x\sec^2(x) - 4\tan(x)}{2x^5}$

c) $7x^3\sec(x)(x\tan(x) + 4)$

d) $\tan(x)\sec(x)$

e) $15\sin(x) + 15\tan(x)\sec(x)$

f) $-2\tan(x)\sec^2(x)$

3.

It is a function within a function (composed). Yes, it applies to other trigonometric functions.

4.

Composed functions cannot be differentiated using the power, product, and quotient rules alone.

CHAPTER 1.6.

1.

a) $-2\sin(x)\cos(x)$

b) $3\sin^2(x)\cos(x)$

c) $16(4x+3)^3$

d) $\frac{34x}{\sqrt{34x^2-24}}$

e) $\frac{8}{3\sqrt[3]{(8x+2)^2}}$

f) $4\sin(x)\cos^2(x) - 2\sin^3(x)$

2.

a) $-8(1+x)^3$

b) $3(4x+5)(2x^2+5x+1)^2$

c) $12(2x+3)^5$

d) $-24\sin(4x)(\cos(3\cos(4x)))$

3.

$R'(t) = 16x^3\frac{dx}{dt} + 14x\frac{dx}{dt} - 5\frac{dx}{dt}$

4.

$\frac{dV}{dt} = 4\pi r^2\frac{dr}{dt}$

CHAPTER 1.7.

1.

a) $-e^{-x}$

b) $-2a^{-x+2}(\ln(a))$

c) $\frac{2}{x\ln(10)}$

d) $\frac{15x^2+2x+3}{x\ln(2)\cdot(5x^2+x+3)}$

e) $\frac{8}{4x+9}$

f) $2x\cdot e^{x^2}$

2.

a) $-\frac{7}{x\ln(4)}$

b) $5\ln(2)\cdot 2^x$

c) $-\frac{3}{x\ln(10)}$

d) $2^{x^2}\cdot 5^{x^2-1}\cdot x\ln(10)$

e) $\frac{3x^2-7}{\left(x^3-7x+5\right)\cdot\ln(4)}$

CHAPTER 1.8.

1.

a) $\frac{2}{4x^2+1}$

b) $\frac{e^x}{\sqrt{1-e^{2x}}}$

c) $\frac{16x(3x^2-1)}{\sqrt{1-(3x^4-2x^2+1)^2}}$

d) $\frac{1}{\sqrt{1-x^2}\cdot\ln(10)\cdot\arcsin(x)}$

e) $\frac{2e^{\arctan(2x)}}{4x^2+1}$

f) $-\frac{6x\ln(5)\cdot 5^{\arccos\left(3x^2\right)}}{\sqrt{1-9x^4}}$

2.

$f'(7)=\frac{1}{2}$

3.

$g'(3)=\frac{1}{16}$

4.

$g'(6)=3$

CHAPTER 1.9.

1.

a)	0	b)	0	c)	$60x$
d)	$200x^3$	e)	$-6x$	f)	0

2.

a) $\frac{7}{x^2\ln(4)}$

b) $5\ln^2(2) \cdot 2^x$

c) $\frac{3}{x^2\ln(10)}$

d) $2^{x^2} \cdot 5^{x^2-1} \ln(10) \cdot \left(x^2\ln(100) + 1\right)$

3.

Inflection points (see 4.2.) require algebraic manipulation of second derivatives (including the second derivative test). Other relationships between quantities in physics also exhibit these relationships (e.g. the second derivative of an object's momentum with respect to time is the force exerted on said object).

4.

Area under the acceleration function (curve) is the velocity, which can also be expressed as: the integral of acceleration with respect to time is velocity. This process can be done simpler using SUVAT equations or inspection and simple geometry (depending on the shape of the area).

PART II

CHAPTER 2.1.

1.

a) $\frac{dy}{dx} = \frac{2x-3y}{3x-2y}$

b) $\frac{-2x-y}{x+3y^2}$

c) $\frac{4x-3x^2y}{x^3+4y^3}$

2.
The instantaneous slope of 1.a) at $(1,0)$ is $\frac{2}{3}$.

3.
The derivative is undefined at the origin (vertical line).

4.
The derivative at the point is equal to 0.

CHAPTER 2.2.

1.
A: deceleration, motion towards the right.
B: constant velocity, motion towards the right.
C: acceleration, motion towards the right.

2.
Jamie's rate of forgetfulness after two days is -16 words/day.

3.
$t = \{12.11, 17.89\}$

4.
$D'(15) = -2.677 \text{ ms}^{-1}$ $D'(25) = 1.996 \text{ ms}^{-1}$

5.
We know that velocity is the time derivative of displacement:

$$\frac{d}{dt}[s] = \frac{d}{dt}\left[ut + \frac{1}{2}at^2\right]$$

$$v = u + at$$

In this instance, initial velocity (u) and acceleration (a) are constant values, and hence the constant multiplication rule applies. After differentiating $v = u + at$ with respect to time, we get $a = a$, which is not useful if we differentiate further.

CHAPTER 2.3.

1.
Tim and Andrew are approaching each other at a rate of (−)5.37 metres per second.

2.
$\frac{dy}{dt} = 112x^6 + 8$

3.
The angle between the bottom of the ladder and the floor is decreasing at a rate of (−)0.125 radians per second or (−)7.16° per second.

4.
After 25 seconds, Jack and Miles are moving away from each other at 1.34 metres per second.

5.
$\frac{dy}{dt} = 61x^2 + 12x$

CHAPTER 2.4.

1.
$g(4.2) \approx 5.6$

2.
$h(-2.9) \approx 1.5$

CHAPTER 2.5.

1.

a) $\lim_{x\to 0} \frac{\tan(x)}{3x + \tan(x)} = \frac{1}{4}$

b) $\lim_{x\to 0} \frac{\sin(x)}{5x + 3\sin(x)} = \frac{1}{8}$

c) $\lim_{x\to \infty} \frac{4x^2-7x}{(\ln(x))^2} = \text{unbounded}$

d) $\lim_{x\to \infty} \frac{-3x^3 + 2x}{4x + 8} = \text{unbounded}$

e) $\lim_{x\to 1} \frac{x\ln(x)}{x^2 - 1} = \frac{1}{2}$

PART III

CHAPTER 3.2.

1.
Critical points of f exist at $x = \{0, 0.874\}$.

2.
A critical point of f exists at $x = 0$.

3.
Critical points of h exist at $x = \pm 0.76$.

4.
f is increasing over the interval $x \in (0, 0.874)$.

5.
f is increasing over the interval $x \in \mathbb{R} \setminus \{0\}$.

6.
h is increasing over the intervals $x \in (-\infty, -0.76) \cup (0.76, \infty)$.

CHAPTER 3.3.

1.
Extrema of f exist at $x = \{-0.5, 1.847\}$.

2.
Local minima of g exist at $x = 0 \pm n \cdot 2\pi \colon n \in \mathbb{Z}$.

3.
A local maximum of h exists at $x = 0.667$.

4.
Extrema of g exist at $x = \{-2.21, 0.176, 1.284\}$.

5.
Local maxima of f exist at $x = \{-4.119, 1.619\}$.

CHAPTER 3.4.

1.
Absolute minimum at $x = 0.64$.
Absolute maximum does not exist.

2.
a) $x \in [0, 2.242]$
b) $x = \{0.366, 0.871, 1.763\}$
c) 3.836 metres
d) 1.763 seconds

3.
Absolute minimum at $x = 1$.
Absolute maximum at $x = 0.199$.

4.
Critical points at $x = \{-0.83, 0.667, 2.971, 4.326, 5.589\}$.
No absolute extrema are observed.

PART IV

CHAPTER 4.1.

1.
Concave upwards when $x \in (0, \infty)$.

2.
Concave downwards when $x \in (\pi, 2\pi)$.

3.
C.U. when $x \in (0, \infty)$; C.D. when $x \in (-\infty, 0)$.

CHAPTER 4.2.

1.
C.D., $x \in (-\infty, -1) \cup (1, \infty)$; C.U., $x \in (-1, 1)$;
Inflection points at $x = -1, 1$.

2.
C.U., $x \in \mathbb{R}$; No inflection points.

3.
C.D., $x > -3$; C.U., $x < -3$; No inflection points.

4.
C.U., $x \in \mathbb{R}$; No inflection points.

CHAPTER 4.3.

1.
Local minimum.

2.
Inconclusive.

3.
a) $t = 38.508$ hours

b) $t = 6.492$ hours

4.
a) $V(x) = x(16 - 2x)(12 - 2x)$ b) $x \in (0, 6)$
c) 194.067cm^3
d) Absolute minimum does not exist over $x \in (0, 6)$, can be infinitely small (must be positive).

CHAPTER 4.4.

1.
a) $x = 6$ b) $(5, -2)$ and $(7, -2)$

2.
a) $x = -1$ b) $(-5, 7)$ and $(3, 7)$

3.
Absolute minimum at $x = 1.732$.
Absolute maximum at $x = -1.732$.

4.
Parallel at $x = 1.26$.

5.
$$\frac{dy}{dx} = \frac{15x^4 - 4xy^2 - 14y}{24x^5 - 3y^2 + 4x^2y + 14x - 6x^5y + 108y^8}$$

CHAPTER 4.5.

1.
$df = \left(2x + 4x^{-1}\right)dx$

2.
$dh = \left(45t^2 + 24t + 3t^{-2}\right)dt$

3.
$\Delta y = 0.158$ $dy = 0.156$

4.
$\Delta f = -0.010$ $\qquad\qquad$ $df = -0.011$

5.
Proof: Obvious. ■

6.
$\frac{1}{5}x^2 + 2x^2 + 7x + C$

7.
$\frac{1}{2}x^4 + \frac{3}{2}x^2 + C$

8.
$2xt + C$

9.
69

10.
$$\int_0^5 4t\,dt + \int_5^{15} 20\,dt + \int_{15}^{25} -2t + 50\,dt$$
$$= 2t^2 \Big|_0^5 + 20t \Big|_5^{15} + (-t^2 + 50t) \Big|_{15}^{25}$$

11.
350 metres

12.
$s(t) = \int v(t)\,dt = \iint a(t)\,dt\,dt$

ACKNOWLEDGEMENTS

This book was a joy to write and compile, and it could not have been possible without some pretty amazing people. Let me embarrass them here for the good of everyone:

Mr. Peter Coyne, who, since walking in his Advanced Mathematics class in Year 10, has allowed me to truly embrace the beauty of mathematics as a subject, including introducing me to matrices, vectors, and of course calculus. He has been an avid collector of the manuscripts and drafts of this piece, many of which are loaded with errors, mistakes, typos, and genuinely poor content which I am sure Sir loved reading in his spare periods as an unconditional subscriber to all future editions of this book.

Mr. Joel Wise, who has, undoubtedly, the best surname in the school; he probably had enough about my inexperience regarding physics and my blatant advertising of this guide. When I presented him with the draft for the first two parts, he looked sternly at me and said, "but this has your name on it…" Yes, Sir because I wrote it. It has been a fun time engaging in the physics classroom talking about vectors and such, and I really do appreciate the help with the calculus portions of 1D, 2D, and 3D motion. Mr. Wise, I understand that the derivative of an exponential function is itself, and a quadratic is fundamentally different, please do not give me a lunchtime detention.

Mr. Dan Rogers, who always says hello to me in our daily-weekly interactions in and around the science centre. He has always kept a lookout for me in my schooling journey, including my questionable decision to pursue architecture as a career back in Years 7 and 8. Sir, thank you so much for making me second-guess myself, and I owe some of the fondest memories to you.

Dr. Terry Byers, who has been mocking me about this book and how I allegedly 'stole' the History Prize on assembly in 2023 even though I dropped history at the end of last year. Sir gave me a 23/25 on learning behaviours because I probably spent too much time on this in class rather than the actual content. Forgive me, but function transformations and factoring quadratics are too boring to do for 50 minutes in one sitting. I also expect this guide to be adopted into the QCAA curriculum, no exceptions.

Dr. Meriet Mikhail, who actually does not know of this textbook's existence. She taught me what a natural logarithm was in our Advanced Science class in Year 10. So far this year, I've just annoyed her by peeping in her classroom and saying hello too many times. Dr. Mikhail was amazing in the year I had her, but I am slightly offended and still cannot comprehend why she did not pick me to be in her senior chemistry class.

Dr. Kristian Saull, who was my awesome Oxbridge tutor for two weeks at Jesus College, the University of Cambridge, during the June-July holidays in 2023. He has prolonged my study of calculus and was probably one of the first people who saw my passion for it and encouraged me to continue pursuing it. I hope that this book has reached you in Cambridge intact, and I sincerely hope you enjoy this read.

Ethan Luk, the 'Ethan' in Example 4.3.2.2., who I have been fortunate enough to meet and play badminton with over the years I have known him and been best friends with. He is a victim to me 'force stuffing' this book into his hands, and now, unbeknownst, has subscribed to all future editions of this book. I trust that he will use this guide well as he ventures into the territory of differential calculus, but I expect he will cruise through the course content. On a side note, want to book a badminton court? Any time.

My maternal grandparents and aunts, who have, since when I was young, supported me on my journey in a foreign environment, and have always accepted me warmly when I take my refuge at their houses every time I return for a 'short' holiday.

My mum, who has always accepted my quirkiness and love for mathematics, amidst the hours of being scolded for being too invested in non-academic pursuits. Thank you, Mum. 爱你。

SPECIAL THANKS

Mr. Grant Smith, director of gifted education (Aquila), former head of Debating, English teacher, who has provided me the opportunities for personal betterment over the five years I have known him. He is possibly the most chill person at school who I can always rely on to get things done.

November 2023 – August 2024

Made in United States
North Haven, CT
04 November 2026

21764045R00095